International Conference on Smart Engineering Materials (ICSEM 2018)

IOP Conference Series: Materials Science and Engineering
Volume 362

Bucharest, Russia
7 – 9 March 2018

ISBN: 978-1-5108-6341-5
ISSN: 1757-8981

Printed from e-media with permission by:

Curran Associates, Inc.
57 Morehouse Lane
Red Hook, NY 12571

Printed by Curran Associates, Inc. (2018)

For permission requests, please contact the Institute of Physics
at the address below.

Institute of Physics
Dirac House, Temple Back
Bristol BS1 6BE UK

Phone: 44 1 17 929 7481
Fax: 44 1 17 920 0979

techtracking@iop.org

Additional copies of this publication are available from:

Curran Associates, Inc.
57 Morehouse Lane
Red Hook, NY 12571 USA
Phone: 845-758-0400
Fax: 845-758-2633
Email: curran@proceedings.com
Web: www.proceedings.com

Table of contents

Volume 362

International Conference on Smart Engineering Materials (ICSEM 2018)

7–9 March 2018, Bucharest, Romania

Accepted papers received: 3 May 2018
Published online: 22 May 2018

Preface

Papers

Material Physics and Film Technology

Building Materials and Structures

Material Chemistry and Chemical Engineering

Mechanical Manufacture and Automation

IOP Conf. Series: Materials Science and Engineering **362** (2018) 011001 doi:10.1088/1757-899X/362/1/011001

PREFACE

2018 International Conference on Smart Engineering Materials (ICSEM 2018) was held in Bucharest, Romania during March 07-09, 2018. ICSEM 2018 was a great success, and academic atmosphere helped participants to exchange scientific and technological ideas and enjoy the material development in the field of smart engineering materials.

ICSEM 2018 provided a real platform for researchers, engineers and academicians to further develop Smart Engineering Materials. Distinguished professors were also invited to deliver keynote speeches relating to the latest developments and trends in their respective fields. The keynote speakers included Prof. Pankaj Vadgama from Queen Mary University of London, UK; Prof. Maria Tomoaia-Cotisel from Babes-Bolyai University of Cluj-Napoca, Romania; Prof. Jordi Arbiol from ICREA & Catalan Institute of Nanoscience and Nanotechnology (ICN2), CSIC and BIST.

This set of conference proceedings presents a selection of papers submitted to the conference from various universities, research institutes and industries. All papers were subjected to peer-review by conference committee members and a panel of international reviewers. The papers were selected based on the quality and relevance to the conference. This volume presents recent advances in the field of Carbonaceus Materials, Nanomaterials and their Arrangements, Two Dimensional Materials, Non Ferrous Metal Material, Smart and Intelligent Materials, Intelligent Materials Systems, Processing Technology for Functional Materials and etc.

We would like to thank all the authors who have contributed to this volume and also to the organizing committee, reviewers, speakers, co-chairpersons, sponsors and all the conference participants for their strong support to ICSEM 2018 to make this a successful event. We look forward to meeting you in ICSEM 2019!

Prof. Pankaj Vadgama
Queen Mary University of London, UK
April 12, 2018

IOP Conf. Series: Materials Science and Engineering **362** (2018) 011002 doi:10.1088/1757-899X/362/1/011002

Organizing Committee

Conference Co-Chairs

Prof. Pankaj Vadgama, Queen Mary University of London, UK

Prof. Maria Tomoaia-Cotisel, Babes-Bolyai University of Cluj-Napoca, Romania

Program Chairs

Prof. Feo V Kusmartsev, Loughborough University, UK

Prof. Jordi Arbiol, ICREA & Catalan Institute of Nanoscience and Nanotechnology (ICN2), CSIC and BIST

Assoc. Prof. Popescu Simona, University Politehnica of Bucharest, Romania

Organizing Chair

Ms. Yang Dan, Hong Kong Chemical, Biological & Environmental Engineering Society, Hong Kong

International Technical Committee

Prof. Yulong Ding, University of Birmingham, U.K.

Prof. Alan Crosky, University of New South Wales , Australia

Prof. Clas Persson, University of Oslo, Norway

Prof. Daniel Chateigner, University of Caen Normandy, France

Prof. Jianbo Wu, Shanghai Jiao Tong University, China

Prof. Kuan-Wei Lee, I-Shou University, Taiwan

Prof. Mohamed Rashad El-Hebeary, Cairo University, Egypt

Assoc. Prof. Kar Seng (Vincent) Teng, Swansea University, UK

Assoc. Prof. Waail Mahmod Lafta, Griffith University, Australia

Assoc. Prof. Yanbin Jiang, University of Science & Technology Beijing, China

Assoc. Prof. Yongchang Liu, University of Science & Technology Beijing, China

Assoc. Prof. Song Lin, Northwestern Polytechnical University, China

Assoc. Prof. Yanfeng Han, Shanghai Jiao Tong University, China

Assoc. Prof. Syarif Junaidi, University of Sharjah, UAE

Dr. Hugh Gong, The University of Manchester, UK

Dr. Srikanth Patala, North Carolina State University, USA

Dr. Laxman Raju Thoutam, University of Minnesota, USA

Dr. Francesca Scalisi, University of Palermo, Italy

IOP Conf. Series: Materials Science and Engineering **362** (2018) 011002 doi:10.1088/1757-899X/362/1/011002

Dr. Santiranjan Shannigrahi, Institute of Materials Research & Engineering, Technology and Research(A*STAR), Singapore

Dr. Huadong Fu, University of Science & Technology Beijing, China

Dr. Liqiang Wang, Shanghai Jiao Tong University, China

Dr. Dingbang Xiong, Shanghai Jiao Tong University, China

Dr. Donghong Wang, Shanghai Jiao Tong University, China

Dr. Yuanfei Han, Shanghai Jiao Tong University, China

Dr. Yujie Chen, Shanghai Jiao Tong University, China

Dr. Yishi Su, Shanghai Jiao Tong University, China

Dr. Jinlong Liu, University of Science & Technology Beijing, China

Dr. Trong-Phuoc Huynh, Can Tho University, Vietnam

IOP Conf. Series: Materials Science and Engineering **362** (2018) 011003 doi:10.1088/1757-899X/362/1/011003

Peer review statement

All papers published in this volume of *IOP Conference Series: Materials Science and Engineering* have been peer reviewed through processes administered by the proceedings Editors. Reviews were conducted by expert referees to the professional and scientific standards expected of a proceedings journal published by IOP Publishing.

IOP Conf. Series: Materials Science and Engineering **362** (2018) 012001 doi:10.1088/1757-899X/362/1/012001

Simulations of Propane and Butane Gas Sensor Based on Pristine Armchair Graphene Nanoribbon

Haroon Rashid[1], Ants Koel[2] and Toomas Rang[3]

1 Ph.D Student at Thomas Johan Seebeck Department of Electronics, Tallinn University of Technology, Estonia
2 Associate Professor at Thomas Johan Seebeck Department of Electronics, Tallinn University of Technology, Estonia
3 Professor at Thomas Johan Seebeck Department of Electronics, Tallinn University of Technology, Estonia
E-mail: haroonrashid199@gmail.com

Abstract. Over the last decade graphene and its derivatives have gained a remarkable place in research field. As silicon technology is approaching to its geometrical limits so there is a need of alternate that can replace it. Graphene has emerged as a potential candidate for future nano-electronics applications due to its exceptional and extraordinary chemical, optical, electrical and mechanical properties. Graphene based sensors have gained significance for a wide range of sensing applications like detection of biomolecules, chemicals and gas molecules. It can be easily used to make electrical contacts and manipulate them according to the requirements as compared to the other nanomaterials. The intention of the work presented in this article is to contribute in this field by simulating a novel and cheap graphene nanoribbon sensor for the household gas leakage detection. QuantumWise Atomistix (ATK) software is used for the simulations of propane and butane gas sensor. Projected device density of the states (PDDOS) and the transmission spectrum of the device in the proximity of gas molecules are calculated and discussed. The change in the electric current through the device in the presence of the gas molecules is used as a gas detection mechanism for the simulated sensor.

1. Introduction

The discovery of graphene has quickly sparked the interest of researchers due to its tremendous promising and unique attributes. The exceptional properties and nontoxic nature of graphene are making it a potential candidates for future electronics applications [1]. Graphene is a single layer of carbon atoms having thickness of 0.35-1.6 nm with a honeycomb lattice structure of sp^2-bonded carbon atoms. It is a two dimensional material with Dirac cones. In graphene lattice, C-C bond length is 1.42 Å and the lattice constant is 2.46 Å. The electronic structure of graphene can be explained by tight-binding approach [2]. Graphene is so called a zero-gap semiconductor [3] which has the possibility to tune its band gap [1]. The unique band structure of graphene give rise to exceptional high electron mobility. It has been reported that the charge carrier transport in the graphene remains ballistic up to 0.3 μm at 300 K. It is highly sensitive to any chemical change in its surrounding environment. Every carbon atom in graphene provides the greatest surface per unit volume and makes mobility of the charge carriers highly sensitive to the adsorbed molecules. Another interesting quality of graphene is its intrinsically low electrical noise due to high quality of crystal lattice. This property makes it highly suitable for ultra high sensitive chemical detection applications [4].

Graphene is being used for high frequency electronic application, field effect transistors, and transparent conductors. Recently graphene based biological and chemical sensor has gained significant importance [5]. Graphene is also emerging as a potential candidate for sensor applications in nano-electronics. Graphene is being used for the detection of gases, biomolecules and heavy metal ions. The adsorption and desorption of gas molecules change the conductivity of the graphene. These gas molecules act as acceptors or donors when they approach the graphene and ultimately they change the conductance of the graphene. The change in the electrical conductance of the graphene nanoribbon is taken as a detection signal for the gas molecules. Molecules of different gases affect the conductance of the graphene differently. So a large number of gases can be detected by using graphene [6]. In this paper pristine armchair graphene nanoribbon based sensor has been simulated for the detection of household alkane group gases like propane (C_3H_8) and butane (C_4H_{10}). To the best of author's knowledge only one closest reference [7] has been found that deals with the usage of the nanotubes for the detection of natural gases that are detected in this paper. In the presence of these gases, change in current through the armchair nanoribbon has been used as a gas detection signal. These simulations are to estimate the device behaviour before physical fabrication experiments of household gas leakage detection sensor.

1.1. Graphene nanoribbons

Different synthetic techniques are being used to synthesize a versatile family of graphene based materials with distinct properties. Graphene nanoribbons belong to the graphene materials family that are highly suitable for sensor applications. The electronic properties of the graphene nanoribbons change from semiconductors to spin-polarized half metals. Graphene nanoribbons are obtained from the unzipping of carbon nanotubes (CNTs) [8]. Depending on the termination pattern of the edge, graphene nanoribbons are divided into two types: armchair edge graphene nanoribbon (AGNR) and zigzag edge graphene nanoribbon (ZGNR). GNRs tend to repeat their geometric sequence to form periodic structures as shown in figure 1 [3].

It has been reported that the bandgap of GNRs changes with the number of carbon atoms constituted by them. It means that the electronic and the magnetic properties of nanoribbons are a function of the length of the ribbons. The bandgap of armchair nanoribbons gradually decreases as the number of carbon atoms increase in the ribbon as shown in figure 2(a). The bandgap of zigzag nanoribbon also tends to decrease gradually for odd and even number of carbon atoms differently in the ribbon as shown in figure 2(b) [9].

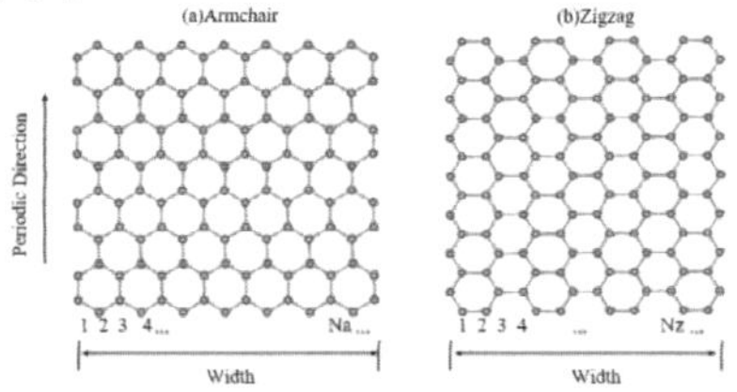

Figure 1. Armchair and zigzag graphene nanoribbons.

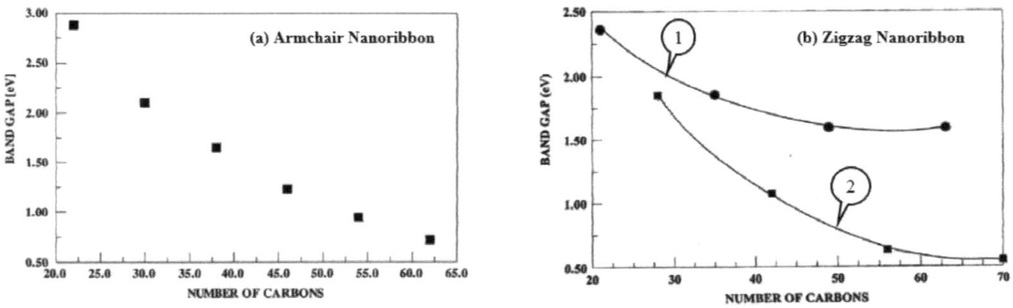

Figure 2. Plots of bandgap energies vs number of carbon atoms for armchair and zigzag graphene nanoribbons where curve 1 is for odd and 2 is for even number of electrons.

IOP Conf. Series: Materials Science and Engineering **362** (2018) 012001 doi:10.1088/1757-899X/362/1/012001

Many theoretical studies such as density functional theory (DFT), tight-binding calculations and many-electron Green's function approach have been done for the investigation of the electronic properties of the graphene nanoribbons. Most of the theoretical studies have been done by using DFT calculations method [3]. In order to enhance the sensing capabilities of graphene, usually it is functionalized by a functional group. Mostly hydroxyl (-OH) and carboxylic (-COOH) groups are used as functional groups for the graphene. In addition, many kinds of chemical moieties such as hydroxyl or alkyl groups can be introduced onto graphene surfaces by adding oxidizing agents. Furthermore, amino groups ($-NH_2$) and sulfonate ($-SO_3$) groups can also be introduced into the graphene. These functional groups act as a chemical handlers to graft different polymers and proteins by creating covalent bonds with them. Hence, in this manner the sensing capabilities of the graphene is increased by covalent bonding [5]. In spite of the stability and the effectivity of covalent strategies, this method is avoided. Because, it unavoidably changes the inherent electronic properties of the graphene by converting sp^2 hybridization of carbon atoms to sp^3 hybridization. So, in order to preserve the intrinsic electronic properties of the graphene; non-covalent modifications to graphene are commonly used [5].

2. Methodology

QuantumWise Atomistix (ATK) software package has been used for the simulations of the gas sensor based on pristine armchair graphene nanoribbon. ATK software package has a graphical user interface called virtual nano lab (VNL). VNL has been used to simulate the structure of graphene based nano sensor as shown in figure 3.

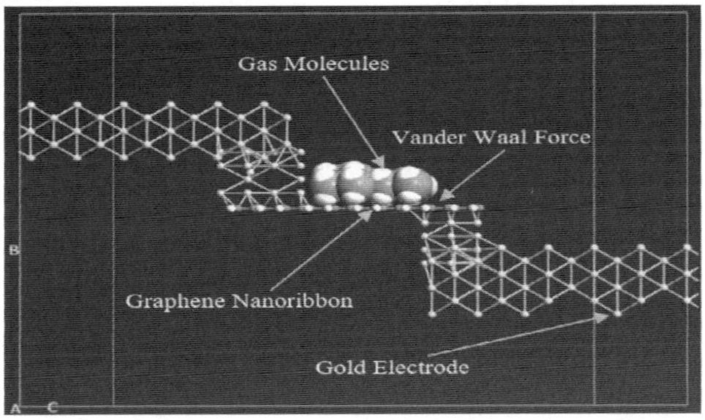

Figure 3. Simulated structure of armchair graphene Nanoribbon based propane and butane gas sensor.

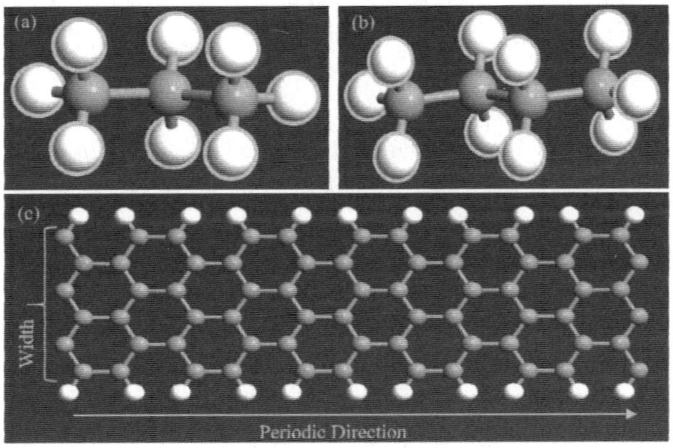

Figure 4. Simulated structure of (a) propane; (b) butane; (c) armchair graphene nanoribbon in ATK.

The Gas molecules are interacting with the armchair graphene nanoribbon by Van der Waal Forces in these simulations. Whereas gold electrodes have been deposited on the nanoribbon for

IOP Conf. Series: Materials Science and Engineering **362** (2018) 012001 doi:10.1088/1757-899X/362/1/012001

characterization of the sensor by applying external voltages. In order to reduce computational time in the simulator; four molecules of the propane gas, four molecules of the butane gas, and two-two molecules of both propane and butane gases are detected by the graphene nanoribbons in three individual experiments. The length of armchair graphene nanoribbon in the simulated device along the periodic direction is 21.31 Å and width is 8.05 Å in figure 4(c). The white balls indicate hydrogen atoms and black balls represent carbon atoms in figure 4(a), (b) and (c).

2.1. Calculation methodology of the ATK-DFT calculator

ATK software has different types of inbuilt calculators to solve Schrödinger equation. ATK-DFT calculator has been used for the simulations presented in this paper. Mathematical formalism [10] used by the ATK-DFT calculator has been explained in this subsection. ATK is a nano-level device simulator that is used to model open and closed systems within the domain of the density functional theory (DFT). Density matrix is the key parameter for the calculations of the Kohn-Sham equations. This density matrix defines the electron density. A non-equilibrium Green's Function (NEGF) is used to calculate the density matrix for open system. Whereas for closed systems diagonalisation of the Kohn-Sham Hamiltonian equation is used to calculate the density matrix. The effective potential set up by the electron density is given by the Hatree, the external potential and the exchange correlation. Then effective potential is used to get the Kohn-Sham Hamiltonian. Electronic structure of a system can be given by one-electron Kohn-Sham Hamiltonian as follows:

$$H1\mathrm{e} = -\frac{\hbar^2}{2m}\nabla^2 + V^{eff}[\mathrm{n}](\mathrm{r}) \tag{1}$$

In equation (1), the first term indicates the kinetic energy of electron whereas the second term indicates the potential energy of electrons moving in the electrical field created by the external potential. Total electron density can be given by, $n = n(r)$. After this one-electron Schrödinger equation is solved to get the eigen-functions of the Kohn-Sham Hamiltonian as:

$$H1\mathrm{e}\Psi_a(r) = \varepsilon_\alpha \Psi_\alpha(r) \tag{2}$$

Equation (2) is called a Kohn-Sham equation with the density functional theory (DFT). This is solved by expanding eigen-functions in a set of basis function, Φ_i:

$$\Psi_\alpha(r) = \Sigma_i c_{\alpha i}\,\Phi_i(r) \tag{3}$$

Now the differential equation can be represented as a matrix equation to determine the expansion coefficients, $c_{\alpha i}$:

$$\Sigma_j H_{ij} c_{\alpha j} = \varepsilon_\alpha \Sigma_j S_{ij} c_{\alpha j} \tag{4}$$

In equation (4), the Hamiltonian matrix H_{ij} and the overlap matrix S_{ij} are given by the multiple integrals with respect to the electron coordinates. Now finally the electron density of many-electron system can be given by the eigenstates of the Kohn-Sham Hamiltonian, thus:

$$n(r) = \Sigma_\alpha f_\alpha \left| \, |\Psi_\alpha(r)|^2 \right| \tag{5}$$

In equation (5), f_α is the occupation of level which has been denoted by α. Then finally the electron density can be given in terms of the density matrix as follows:

$$n(r) = \Sigma_{ij} D_{ij} \Phi_i(r)\Phi_j(r) \tag{6}$$

3. Results and discussion

This section contains the results and discussion of the simulations of gas sensor. The transmission spectrum and projected density of states of the device at zero bias condition are given and discussed. Furthermore, the graphs for the change in the current through the device in the proximity of the propane and the butane gases are also discussed in this section.

3.1. Projected density of the states of the device

The projected density of the states (PDOS) versus energy of the graphene nanoribbon based gas sensor is shown in figure 5. It can be clearly observed from these plots that many energy states are available to be occupied below and above the fermi level. The Projected density of the states are indicating the presence of many energy states around the fermi level. Comparison of the density of the states (DOS) with respect to corresponding energy values of the device reveals a considerable difference in DOS.

Figure 5(a) is the density of the state of the sensor in the absence of any gas while (b) is the DOS in the presence of the propane gas. It can be seen in figure 5(a) that there are three sharp peaks appearing at the energy levels of approximately 1.2 eV, 1.7 eV and 2.4 eV. Whereas in figure 5(b) the presence of propane gas has influenced the DOS of the device. Hence, there is only one sharp peak is appearing at the energy level of 1.2eV in this graph as compared to figure 5(a).

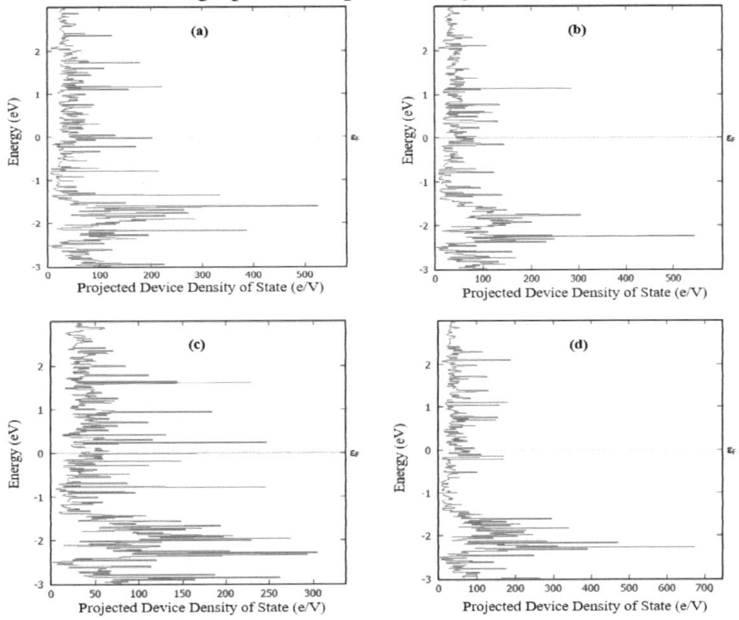

Figure 5. Projected density of the states of the simulated sensor vs. energy where (a) graphene sensor without gas; (b) sensor with propane gas; (c) sensor with the butane; (d) sensor with propane and butane gas.

Comparison of the DOS of the device in the presence of butane gas in figure 5(c) with the sensor in the absence of gas in figure 5(a) reveals that a lots of energy states have been added by the butane gas. Many sharp peaks of the density can be seen above and blow the fermi level as compared to figure 5 (a) and (b). Finally the density of the states of the device under the joint influence of the propane and the butane gases are shown in figure 1(d). It can be observed in this graph that the energy states have been modified differently as compared to figure 1(b) and (c). These modifications in the density of the states by these gases have a strong influence on the electrical conduction properties of graphene based gas sensor.

3.2. Transmission spectrum of the device

Transmission spectrum of the device is calculated to analyse the contribution of different energy levels in the total transmission of the charge carriers through the device [11]. The plots of the transmission as a function of the energy are shown in figure 6 and 7. It can be seen in figure 6 that for the same energy levels, the value of the transmission is different for the different gases. In the absence of any gas at energy level of E=-1.32 eV, the value of transmission is 0.606 for the sensor as shown in figure 6(a). Whereas for the same energy level, the value of the transmission for the device in the proximity of the propane gas is 0.0046, as shown in figure 6(b).

Similarly in figure 6(c); the value of transmission as a function of the same energy level is 0.00034 in the presence of butane gas. This value of transmission is quite different from the values that are shown in figure 6 (a) and (b). At E= -1.32eV, in the presence of the propane and the butane gases the transmission value is 0.0486 as shown in figure 6(d). So, it is clear from these results that the presence of different gases alters the value of the transmission of the charge carrier through the device for the same energy levels.

IOP Conf. Series: Materials Science and Engineering **362** (2018) 012001 doi:10.1088/1757-899X/362/1/012001

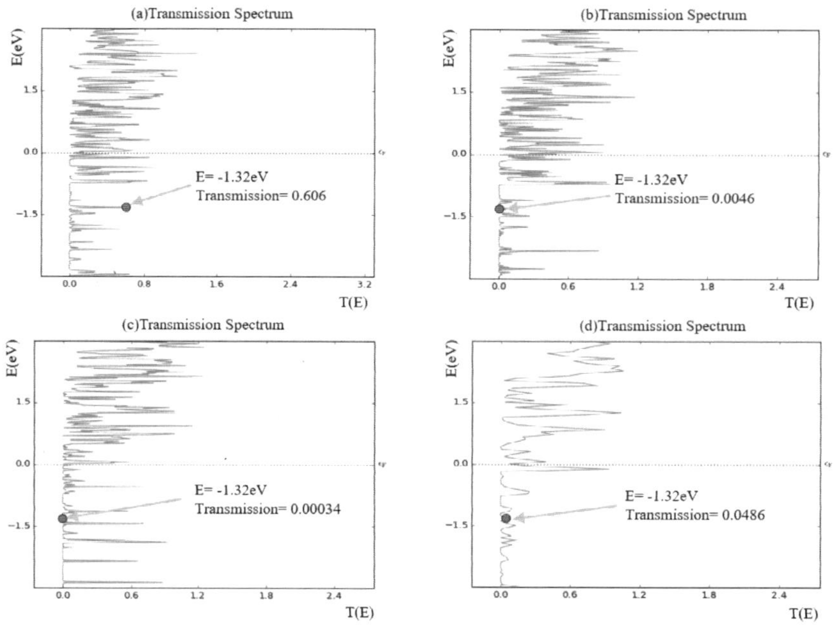

Figure 6. Transmission spectrum as a function of energy of simulated sensor at different energy levels (a) in the absence of gas; (b) with propane gas; (c) with the butane gas; (d) with the propane and butane gas.

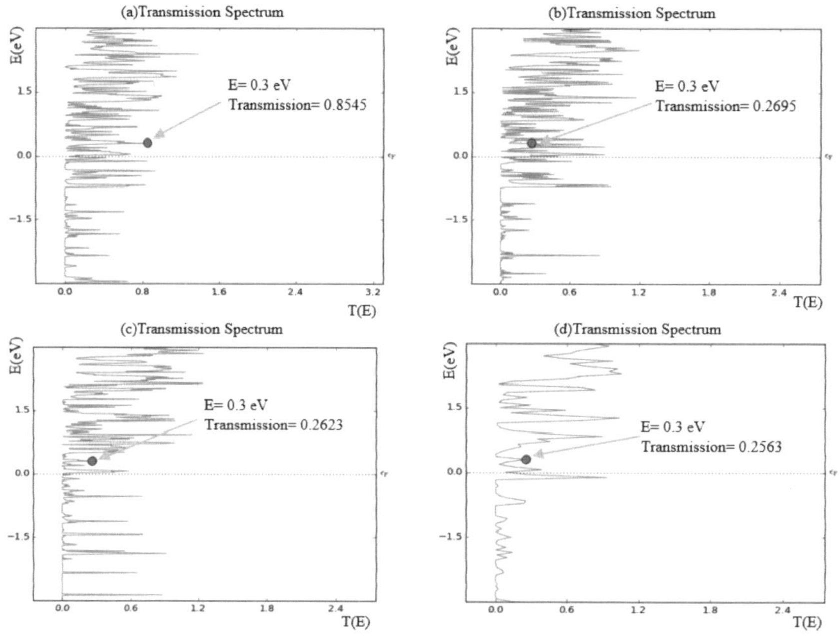

Figure 7. Transmission spectrum as a function of energy of simulated sensor at different energy levels (a) in the absence of gas; (b) with the propane gas; (c) with the butane gas; (d) with propane and butane gas.

Furthermore, the transmission spectrum at different energy levels of the same device in the presence of the same gases is also presented in figure 7. It can be observed in figure 7 (a) that at energy level E= 0.3 eV, the value of transmission is 0.8545 in the absence of any gas. While, for the same energy level for the propane gas the value of transmission is 0.2695 in figure 7(b). For the same energy level, in the presence of the butane gas transmission is 0.2623 as shown in figure 7(c). Whereas the transmission is 0.2563 at the same energy level in the presence of the butane and the propane gases as shown in figure 7(d).

3.3. Current-voltage curves of the device

The electrical conductivity of the majority of the graphene based gas sensors changes in the presence of foreign gas molecules. This change in the conductivity is used as a gas detection mechanism in the graphene based gas sensors [4]. The same principle has been used to develop the propane and the butane gas sensor based on the armchair graphene nanoribbon in our simulations. DC bias voltages are applied at the electrodes of the simulated sensor to obtain the current-voltage (IV) curves. These simulations are done in the absence of any gas, in the presence of the propane gas, in the presence of the butane gas and in the presence of both propane & butane gases to obtain different IV-curves. These IV-curves are shown in figure 8. The Curve 1 in figure 8 is representing the current through the simulated device when there was no gas in the proximity of the sensor. It can be observed that the device is showing almost resistive behaviour in all cases. Curve 1 is revealing that the device starts to conduct immediately even at a very low bias voltage. In the absence of any gas, the IV-curve of the device is almost linear with the high current flowing through the device.

Curve 2 in figure 8 is obtained in the presence of the propane gas molecules. It can be observed in this curve that the presence of the propane gas molecules has reduced the current through the device as compared to curve 1. The adsorbed gas molecules act as donors or acceptors for the graphene and change its conductivity [4]. Graphene shows ultrahigh sensitivity in the presence of the foreign gases because the pristine graphene is inherently a low noise material. The presence of gases near the graphene introduce electrical noise and disturbs its electronic cloud. As the conductivity of the charge carriers through the device depends on the number of charge carriers and their mobility. So, the presence of the propane gas may have changed the number of the charge carriers or their mobility in the graphene nanoribbon. Due to which Curve 2 is showing a decrease in current through the device. The change in charge carriers concentration, changes the current through the device and acts as a governing mechanism for the conductivity of the device [4, 5].

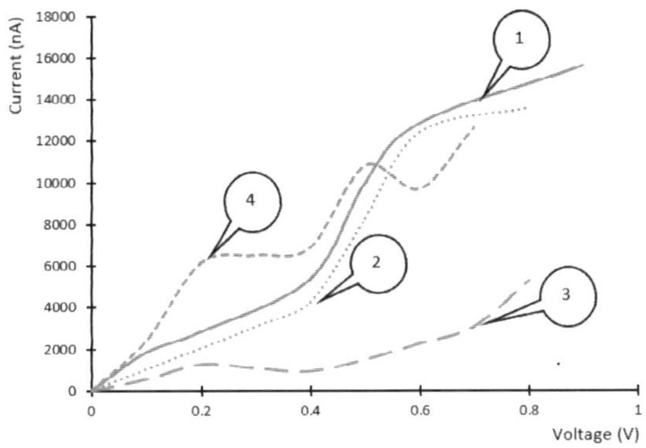

Figure 8. Current-voltage curves of simulated armchair graphene Nanoribbon based gas sensor where curve (1) in the absence of gas; (2) with propane gas; (3) with butane gas; (4) with propane and butane gas.

Furthermore, in the presence of the butane gas a significant reduction in the current through the sensor is observed as shown in Curve 3 in figure 8. The possible reason of this reduction in the current can be the decrease in the number of the majority charge carriers and their scattering by the butane gas molecules [8]. An interesting behaviour of IV-curve is observed in Curve 4 in figure 8 when the both propane & butane gases were in the proximity of the device. It can be observed from this curve that the current through the device has increased significantly. In addition to the electric current, the nonlinearities in this IV-curve 4 are also increased. The reason of this increase in the electric current up to a certain voltage level and then decrease could be due to the joint influence of the scattering process provoked by these both gases.

4. Conclusion

The objective of the current work is to simulate the graphene nanoribbon based sensor for the leakage detection of household gases and estimate device behaviour before physical fabrication. The nanoscale simulations of the gas sensor are done successfully with the QuantumWise Atomistix software. The results of these simulations revealed that the graphene is highly sensitive to the change in its surrounding environment. The presence of the gases in the proximity of the graphene device strongly influenced its density of the states, the transmission spectrum and the electric current, when propane and butane gases were in its proximity. The reasons of this change in the conductivity are an increase or decrease in the charge carriers and/or their mobility caused by the adsorbed gas molecules on the surface of graphene. A decrease in the current through the sensor is observed when propane and butane gases are interacting with it individually. But when both gases are interacting simultaneously with the graphene, an increase in the current and the nonlinearities are observed. This change in the conductivity through the device can be used as a gas detection signal for sensor applications. Although the graphene is a potential candidate for sensor applications in future electronics, but there are still a large number of technical challenges that need to be addressed. These challenges are the selectivity of graphene to detect the specific gases, reliability and the cost of the device. Further investigations are required to improve the performance of our simulated graphene based gas sensor. For this purpose, more than one detection mechanisms can be used simultaneously (i.e. change in optical and electrical properties) to avoid errors in the sensor readings due to the contamination of undesired particles and the effect of the temperature in the sensing environment.

References

[1] Zhang Y *et al.* Direct observation of a widely tunable bandgap in bilayer graphene 2009 *Nature* **459** 820-23
[2] Wang J Deng S and Liu Z The rare two-dimensional materials with Dirac cones 2015 *Natl. Sci. Rev.* **2** 22-39
[3] Khan E Li Z and Yang J *Physics and Applications of Graphene – Theory* (InTech) pp 332-348
[4] Yavari F Koratkar N Graphene based chemical sensors 2012 *J. Phys. Chem. Lett.* **3** 1746-53
[5] Liu Y Dong X and Chen P Biological and chemical sensors based on graphene materials 2012 *Chem. Soc. Rev.* **41** 2283-2307
[6] Kuila T Bose S Khanra P Mishra A K Kim N H and Lee J H Recent advances in graphene-based biosensors 2011 *Biosens. Bioelectron.* **26** 4637-48
[7] Keshtkar S Rashidi A Kooti M Development of tin dioxide quantum dots/multi-walled carbon Nanotubes and tin dioxide quantum dots/carbon nanohorns nanohybrids as low temperatures Natural gas sensors 2017 *Cera. Intl* ***43*** 14326-1433
[8] Erol O Uyan I Hatip M Yilmaz C Tekinay AB and Guler O Recent advances in bioactive 1D and 2D carbon nanomaterials for biomedical applications 2016 *Nanomed.* 1-22
[9] Owens F J Electronic and magnetic properties of armchair and zigzag graphene nanoribbons 2008 *J. Chem. Phys.* **126**
[10] QuantumWise User Manual: ATK-DFT Calculator. Internet: https://docs.quantumwise.com/manuals/ATKDFT.html [Nov. 20, 2017]
[11] QuantumWise User Manual: ATK Transport Calculations. Internet: https://docs.quantumwise.com/tutorials/atk_transport_calculations/atk_transport_calculations.html [Nov. 24, 2017]

Acknowledgments

This research was supported by the Estonian Research Council through the Institutional Research Project IUT19-11, by the Estonian Research Council through the Institutional Research Project PUT1435, and by the Horizon 2020 ERA-chair Grant "Cognitive Electronics COEL" —H2020-WIDESPREAD-2014-2 (Agreement number: 668995; project TTU code VFP15051).

IOP Conf. Series: Materials Science and Engineering **362** (2018) 012002 doi:10.1088/1757-899X/362/1/012002

Design and Optimization of AlN based RF MEMS Switches

Mehadi Hasan Ziko[1] and Ants Koel[2]

[1] Ph.D student at Thomas Johann Seebeck Department of Electronics, Tallinn University of Technology, Estonia
[2] Associate Professor at Thomas Johann Seebeck Department of Electronics, Tallinn University of Technology, Estonia
E-mail: mehadi.ziko@ttu.ee

Abstract. Radio frequency microelectromechanical system (RF MEMS) switch technology might have potential to replace the semiconductor technology in future communication systems as well as communication satellites, wireless and mobile phones. This study is to explore the possibilities of RF MEMS switch design and optimization with aluminium nitride (AlN) thin film as the piezoelectric actuation material. Achieving low actuation voltage and high contact force with optimal geometry using the principle of piezoelectric effect is the main motivation for this research. Analytical and numerical modelling of single beam type RF MEMS switch used to analyse the design parameters and optimize them for the minimum actuation voltage and high contact force. An analytical model using isotropic AlN material properties used to obtain the optimal parameters. The optimized geometry of the device length, width and thickness are 2000 μm, 500 μm and 0.6 μm respectively obtained for the single beam RF MEMS switch. Low actuation voltage and high contact force with optimal geometry are less than 2 Vand 100 μN obtained by analytical analysis. Additionally, the single beam RF MEMS switch are optimized and validated by comparing the analytical and finite element modelling (FEM) analysis.

1. Introduction

The RF MEMS switch is the most important MEMS device in the RF/microwave field. A RF MEMS switch is a switching device used in the RF range, which is fabricated using the micromachining technology. More flexible, light weight and low power wireless systems are the usual demand for RF communication that can drastically reduce manufacturing cost, size, weight and improve performance [1] and battery life. Commercialization of RF MEMS switches has several challenges due to high voltage and high power required for the available RF MEMS switches. Due to high voltage requirements, the functional benefit from the RF MEMS switches has been hampered. This power related problems are minimized and various requirements and issues must compare with design parameters to get good performance by the piezoelectric material based RF MEMS switch. Piezoelectric actuation principles seems to be advantageous over the electrostatic actuation for MEMS applications because of the operation with high force, fast speed response with low actuation voltage and easy control, good operational repeatability, high operational efficiency, possibility of static and vibration actuation, lack of dust, radiation and temperature harmfulness features [2, 3]. Various research groups and commercial companies intended to do research on piezoelectric RF MEMS switches based on PZT (Lead zirconate titanate) materials over the last few years. However, for good performance RF MEMS switch, it is necessary to improve the actuation voltage and contact force, which depends on the piezoelectric coefficients of a piezoelectric material. PZT has high piezoelectric coefficients that can increase the contact force and the actuator deflections. Nevertheless, the PZT are not compatible with CMOS. So, based on literature study focusing on various actuation mechanisms and analytical calculation it is preferred that the piezoelectric AlN based RF MEMS switch can fulfill good RF performance requirements. Theoretical study has been done in these recent research for electrostatic [6,7], PZT [10, 5] and AlN [3, 4] based RF switches. Considering the RF performance and technological relationship, it is observed that, AlN based RF MEMS switch would be best compare to the electrostatic and PZT based RF MEMS switches.

2. Analytical design and optimization of RF MEMS switches

The analytical solution in this section used the Euler-Bernoulli beam theory because shear deformation is negligible for a long beam. A variable cross-section is used e.g. to reduce the actuation voltage and increase the contact force for a piezoelectric actuated RF MEMS switch. Therefore, several different design approaches are investigated using an analytical design technique based on 3-D piezoelectric single beam RF MEMS switches. Because of this hypothesis, strain distribution within the piezoelectric film to be homogeneous. The expression for the average induced voltage and beam displacement for a unimorph multilayer beam, subjected to a static load are discussed and studied in this work. The analytical methods were implemented in MathCad, where a number of parameters can be differed at once. Utilizing this approach, the effect of given parameter could be better comprehended allowing for better optimization of a structure. With a suitably polarized, it is demonstrate the configuration of a basic piezoelectric beam thin-film actuator switch. Here the length and width of the piezo-electric layer and the substrate layer are same in dimension with different thicknesses. The beam deflection is depending on the polarization of the piezoelectric film and actuation voltage. The analysis of such switch used to determine the displacement that can be achieved by the piezoelectric actuation under some conditions. These conditions are studied in this analysis.

Area of the beam substrate layer and piezoelectric layer are expressed by the following equations:

$$A_b = W_b L_b \tag{1}$$

$$A_p = W_b L_p \tag{2}$$

Here A_b and A_p are the area of the substrate and piezoelectric layers respectively. The length and width of the beam substrate and piezoelectric layers are expressed by the L_b, L_p and W_b, W_p respectively. The considered beam moment of inertia for the substrate layer I_b and piezoelectric layer I_p are expressed by the following equations:

$$I_b = W_b t_b^3 / 12 \tag{3}$$

$$I_p = W_p t_p^3 / 12 \tag{4}$$

Where, t_b and t_p are the substrate and piezoelectric layers thickness of the single beam RF MEMS switch.

In the piezoelectric single beam RF MEMS switch, the strain distribution on the piezoelectric layer is dominated by the deflection pattern of the substrate. As a result, the free piezoelectric strain ε_0 is linearly distributed over the substrate, which can be expressed by the equation 5 [8].

$$\varepsilon_0 = V d_{31} / t_p \tag{5}$$

The AlN base single beam RF MEMS switch piezoelectric bending moments M_s can be describe as [9].

$$M_S = \frac{1}{1+\frac{E_p I_p}{E_b I_b}} \frac{t_b + t_p}{2} * \varepsilon_0 / \left[\left(\frac{1}{E_p A_p}\right) + \left(\frac{1}{E_b A_b}\right)\right] + \left[\left(\frac{t_b^2}{4E_b I_b}\right) + \left(\frac{t_p^2}{4E_p I_p}\right)\right] + \left[\left(\frac{t_b t_p}{4E_p I_p}\right) - \left(\frac{t_b^2}{4E_b I_b}\right)\right] + \left[\frac{(r t_b - t_p)}{(1+r) t_b}\right] \tag{6}$$

Here E_b and E_p are the young modulus of substrate and piezoelectric layer respectively. The above equations that are used to calculate the static deflection of a single beam RF MEMS switch actuated by a piezoelectric actuation can be calculated by:

$$\mathrm{Disp} = M_S L_b^2 / 2E_b I_b \tag{7}$$

The displacement analysis of single beam RF MEMS switches are studied by equation 7. Figure 1 shows single beam piezoelectric RF MEMS switch displacement, which increase as the beam length increase with the substrate layer thickness. But, in contrast to piezoelectric layer of the RF MEMS switches, higher displacement obtained for the thinner piezoelectric layer over the substrate layer shown in figure 2. Displacement as a function of applied voltage correspond to beam length also studied and observed that, larger displacement is obtained for the higher voltage for longer beam length. This analytical analysis represents the behaviour of various geometrical influences and effect of voltage loading on displacement for the single beam RF MEMS switches in piezoelectric actuation. So, multilayer beam actuated by the piezoelectric layer RF MEMS switches significantly depends upon the original length of the beam, substrate thickness, thickness of the piezoelectric layer and applied voltage.

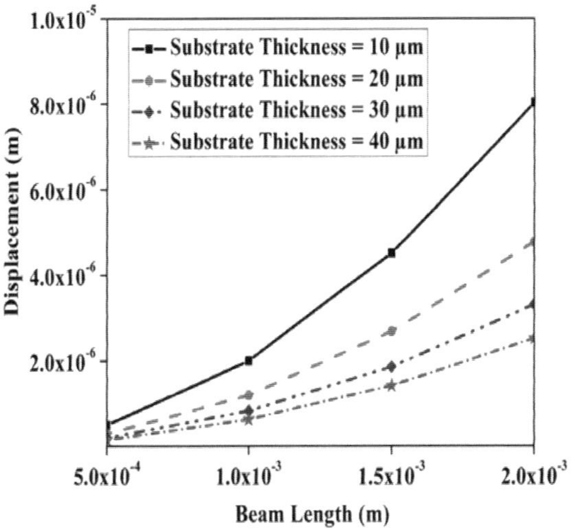

Figure 1. Single beam RF MEMS switches displacement relationship with length and substrate thickness.

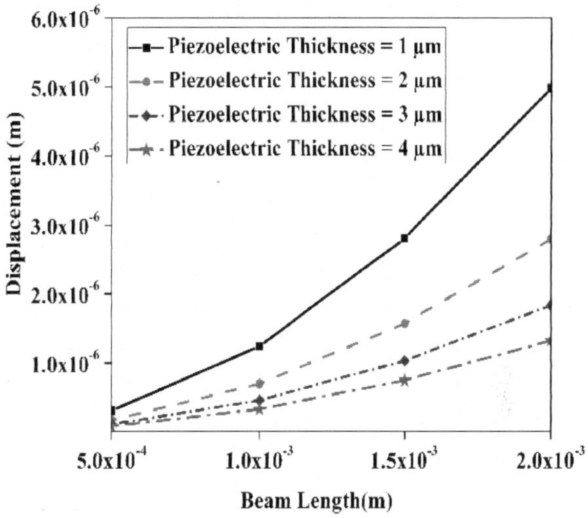

Figure 2. Single beam RF MEMS switches displacement relationship with length and piezoelectric layer thickness.

2.1. Analytical voltage analysis

The RF MEMS switch performance are depends on the actuation voltage of the device. So it is necessary to minimize the voltage to get good RF performance of MEMS switches. In order to lower the actuation voltage of the MEMS switch, three main different rules can be followed: 1) designing a structure with a low spring constant; 2) decreasing the gap between the membrane and the bottom electrode; and 3) increasing the effective actuation area [1]. The first one is the most flexible, because the design of the springs constant does not considerably impact the size and weight of the RF performance and its circuits. In the second case, the isolation associated with the RF signal, which depends on the gap size. And the third one is the effective area of the piezoelectric actuation electrodes, which is limited with regard to the total area of the device. Most of these parameters are directly linked to each other. As an example, if it decreases the gap or raise the area of the piezoelectric and substrate layer, this results increase of off capacitance, leading to a poor isolation. So, there is a certain limit to decrease the gap size. A factor (K) is required to estimate the actuation voltage for the single beam RF MEMS switch that include the geometrical and material constants.

$$K = \frac{\frac{t_b}{2t_p}}{\frac{1}{E_b(W_b t_b)} + \frac{t_b^2 + t_p t_b}{4E_b I_b} + \frac{1}{E_p W_b t_p}} \tag{8}$$

The actuation voltage are explain by the equation 8 and equation 9 [9].

$$V = 2E_b I_b G / K d_{31} L_b^2 \tag{9}$$

2.2. Analytical Contact force analysis

The contact force analysis is necessary for the control and reduced the contact resistance that is desirable for the good performance RF MEMS switches. The gap size is also an important parameter for the contact force analysis. The initial gap "G" is usually fixed by the required isolation loss between the switch contact electrodes. It is investigated that, the contact force is inversely proportional to the gap size. The Gap size can be calculated by the equation 10 [8].

$$G = \frac{M_e}{2(EI)_e} L_b^2 - \frac{F_c}{3(EI)_e} L_b^3 \tag{10}$$

So, the contact force analysis for a single beam RF MEMS switch structure can be derived as:

$$F_c = \left[\frac{M_e L_b^2}{2(EI)_e} - G\right] \frac{3(EI)_e}{L_b^3} \tag{11}$$

Figure 3 illustrate that, after a certain substrate thickness, the contact force will be decreasing and every 100 μm substrate width increase, the contact force will increase almost double. It can also demonstrated that, the contact force decreases after a certain substrate layer thickness. So, the contact force analysis of the geometrical parameters for the RF MEMS switch is really difficult. To get the maximum contact force it is necessary to optimize the geometrical parameters.

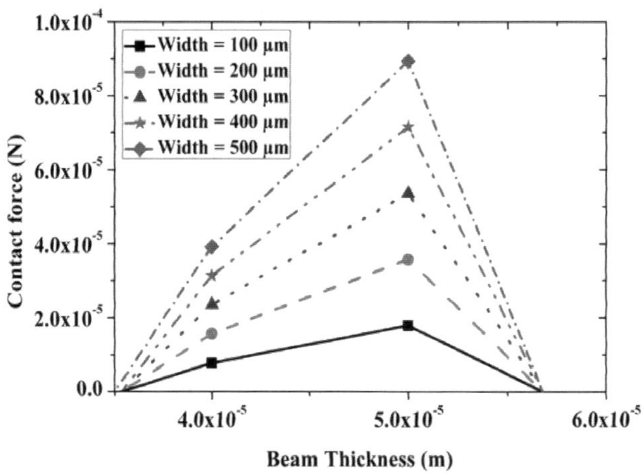

Figure 3. Contact force relationship with beam width and substrate layer thickness for single beam RF MEMS switches.

3. Optimization of geometry for minimum voltage and high contact force

3.1. Minimum voltage optimization

All geometrical parameters must fulfill the fabrication requirements, which are the major considerations of the geometrical design for RF MEMS switches. This research will introduce the

IOP Conf. Series: Materials Science and Engineering **362** (2018) 012002 doi:10.1088/1757-899X/362/1/012002

various analytical analysis to minimize the actuation voltage by finding the optimal geometrical parameters. From the figure 4 it can be illustrated that, the higher substrate thickness needs higher voltage for actuation. It can be stated that, the thickness of the substrate layer is more impacting on the voltage than the width of the substrate layer size. This validates the previous discussion (selecting the substrate thickness). From the analysis, it also observed that single beam piezoelectric RF MEMS switch actuation voltage will decreased as the beam length increases. The actuation voltage reduces to less than 1 Volt is obtained from the beam width of 500 μm with the length of 2000 μm for a certain thickness of the beam. After that, as the substrate layer thickness of the RF MEMS switch increase, the actuation voltage will also increase. As discussed before, the substrate layer thickness should be sufficiently thin for the piezoelectric actuation. Therefore, the beam length of 2000 μm with the substrate thickness of the 50 μm are considered as optimal parameters for the minimum actuation voltage.

These geometrical parameters will be suitable for the AlN based piezoelectric actuation and the minimum voltage of less than 2 Volt is considered for the optimum actuation voltage. It is also satisfy the increasing effective actuation area for minimizing the actuation voltage. As it is previously discussed that, decreasing the gap between the beam and the bottom electrode which lead to minimize the voltage. This statement validated by the figure 5. Higher gap size needs higher voltage to actuate the single beam RF MEMS switches. The gap size of 500 nm considered as the optimal gap for the minimum actuation voltage.

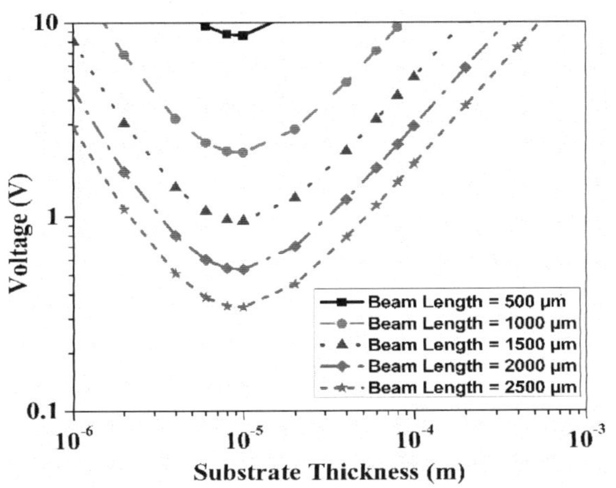

Figure 4. Single beam RF MEMS switch actuation voltage and substrate thickness relationship by changing beam length.

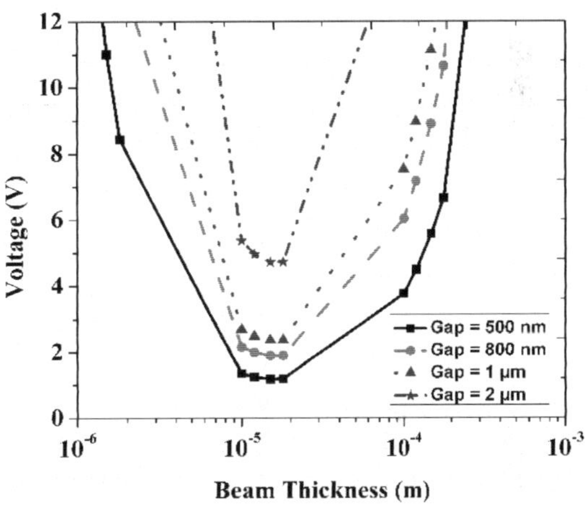

Figure 5. Single beam RF MEMS switch actuation voltage relationship with substrate layer thickness and gap size between the beam and the bottom electrode.

Analysis also done for the optimization of piezoelectric coefficient. From this analysis it can be explain that, the minimum voltage also depends on the piezoelectric coefficient along with the geometrical cross-section. RF MEMS switch actuation voltage can be minimized by increase the piezoelectric coefficient. But it is limited by the some certain value of piezoelectric coefficient (d_{31}) which can be obtained from the material selection. After a certain limit, if the piezoelectric coefficient increases, the minimum voltage will not be decreased significantly. It can be described that, the voltage of the RF MEMS switches depend on the piezoelectric coefficient (d_{31}), which could be determined during the deposition process. So, after the analytical design analysis, the optimal piezoelectric coefficient is 2.6 (pm/V) considered for this single beam RF MEMS switches study.

3.2. High contact force optimization

In this analysis, previous analytical design experience used to select the geometrical parameters. From the figure 6 it can be describe that, the contact force 100 - 200 μN is obtained for the optimal geometry. This range of contact force in RF MEMS switch is desirable for the good RF performance [1].

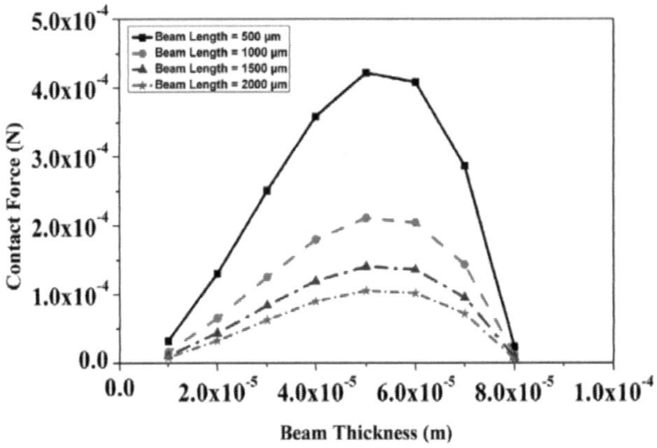

Figure 6. Contact force relationship with substrate thickness for single beam RF MEMS switch.

Here the gap size between beam and electrode is 500 nm considered. For this case, the piezoelectric co-efficient are considered as 2.6 pm/V and piezoelectric thickness kept constant as 0.6 μm. Based on the presented equation 11, optimization of the contact force analysis of single beam RF MEMS switches are driven by piezoelectric actuator and maximize by given isotropic materials, initial gap and actuation voltage. It is observed that contact force is maximized when the single beam RF MEMS switch length allows a free tip deflection and it is mainly depends on the length and the gap size of the single beam RF MEMS switch.

4. Conformation of modelling tool for RF MEMS switches

Analytical models are extremely useful for a preliminary assessment of RF MEMS switches at the design stage. The analytical solution has an error, because it is an approximate solution of theoretical equations. The solid element type will be a better choice as long as the RF MEMS switch is primarily under bending which were the conditions of the analysis in this section. The majority of the modelling using the analytical methods presented in the previous section. FEM simulations are performed for the electromechanical modelling using the ANSYS platform as conformation modelling. Finally, the optimized analytical results for single beam are used as the reference to check the performance of the FE simulation model. This way of analysis will also validate the analytical result.

4.1. Single beam RF MEMS switch analytical and numerical simulation comparisons

To estimate the piezoelectric single beam RF MEMS switch displacement it is considered the values and proportions of the optimized beam. The analytical analysis and FEM ANSYS solution points of displacement are shown below in subsequent figures. The material properties considered in the ANSYS models are all anisotropic. The anisotropic material properties of Si and SiO_2 are taken from the referred paper [11].

The boundary conditions applied to single beam type piezo-electrically actuated RF MEMS switch are shown in the figure 7 (a) and RF MEMS switch FEM model shown in the figure 7 (b). The figure 8 describes the behaviour of voltage loading in the case of the single beam RF MEMS switch by analytically and numerically for piezo-electric actuation.

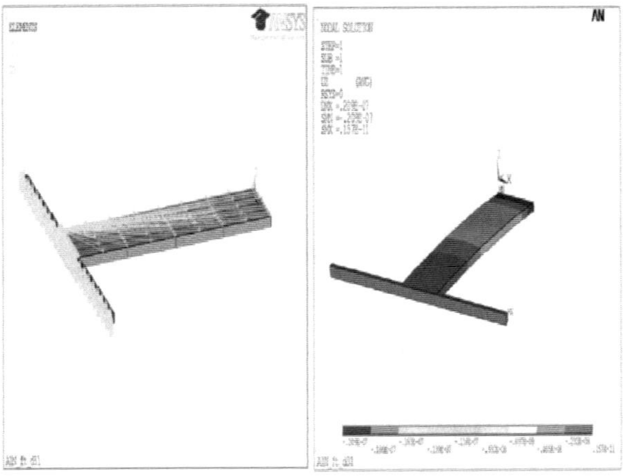

(a) (b)

Figure 7. Single beam RF MEMS switch (a) boundary conditions applied to piezoelectric actuated RF MEMS switch (b) RF MEMS switch FEM model.

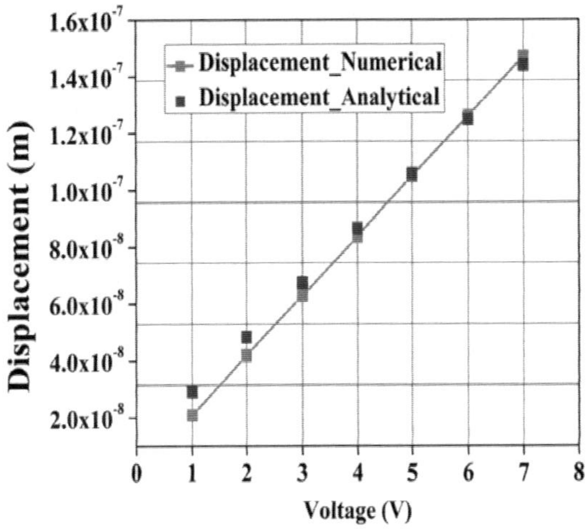

Figure 8. Single beam RF MEMS switch analytical and numerical analysis comparison between displacement and voltage.

Displacement is linearly increased as the voltage increased. Larger displacement is obtained by applying a higher voltage for ANSYS and analytical simulation. The FEM and analytical simulation displacement for the single beam RF MEMS switch are very good agreement. Displacement investigation as a function of beam length, width are plotted shown in the figure 9 and figure 10 respectively using analytical and numerical (FEM ANSYS) solution by applying 2 voltage. It can be seen that single beam piezoelectric RF MEMS switch displacement will increase as the beam length and width increase.

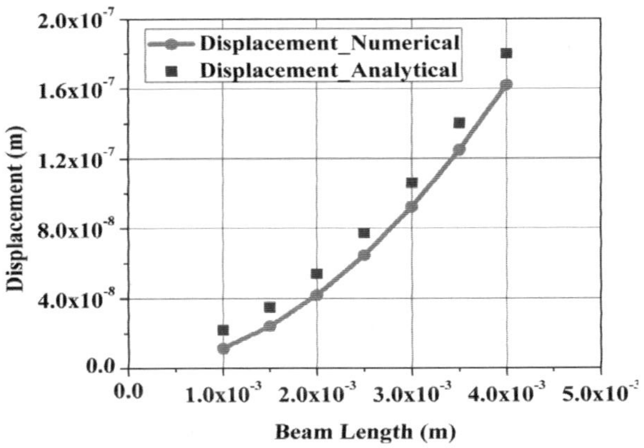

Figure 9. Single beam RF MEMS switch analytical and numerical analysis comparison between displacement and length.

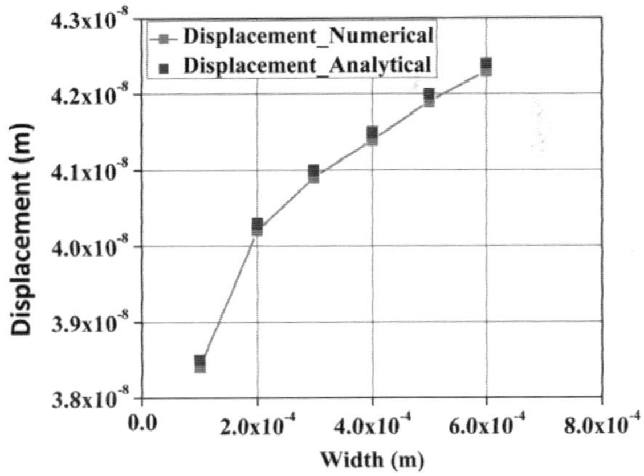

Figure 10. Single beam RF MEMS switch analytical and numerical analysis comparison between displacement and width.

It can be seen that, single beam piezoelectric RF MEMS switch displacement will increase as the beam length and width increase. From the figure 11 it is observed that the displacement is linearly decrease as the thickness of the piezoelectric layer increases for the case of ANSYS simulation. But in contrast of piezoelectric layer thickness, displacement decrease exponentially with the increasing piezoelectric layer thickness. This is because, thinner piezoelectric layer of RF MEMS switch needs more voltage for the displacement that is discussed in the analytical simulation. The possibility of this reason is that, stress becomes increasingly important as the thickness of the piezoelectric film increases relative to the substrate layer.

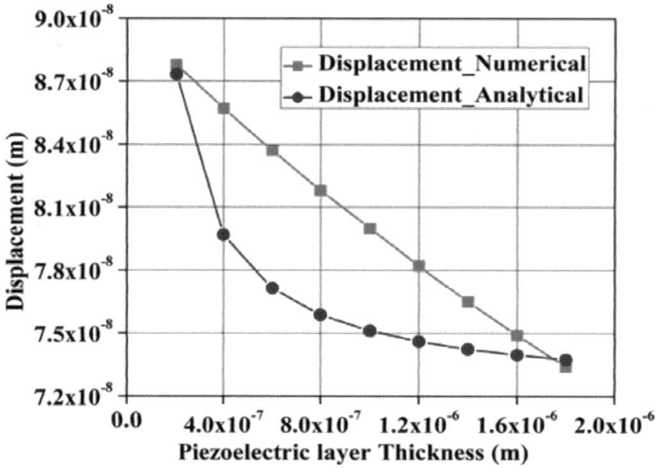

Figure 11. Single beam RF MEMS switch analytical and numerical analysis comparison between displacement and piezo-electric layer thickness.

Single beam RF MEMS switch analytical and numerical analysis comparison for displacement have slight difference in the results. This can be attributed to the fact that, the analytical solutions are modelled by using isotropic material properties and the FEM model is based on anisotropic material properties. There is also one limitation that, the FEM model having a boundary condition where the analytical model do not have any boundary conditions.

5. Conclusions

The analytical model for the single beam RF MEMS switch is developed starting with the basic equations. Several assumptions are made in developing the analytical models that are best suited for behaviour of the single beam RF MEMS switches. Single beam type RF MEMS switches actuation voltage and contact force are analysed by the appropriate equations. Theoretical analysis of the design parameters are optimized for the minimum actuation voltage and high contact force. The optimized beam length and width of 2000 μm× 500 μm are obtained from the analytical analysis. Piezoelectric film thickness of 0.6 μm should be controlled at the smallest dimension considering the biggest electric gradient that the material can sustain. The optimized gap size and piezoelectric coefficient of 500 nm and 2.6 pm/V are obtained for the efficient and reliable operation of the switch. Using these methods, the parameter dependency of the different design characteristics are investigated. The optimized minimum actuation voltage and high contact force are less than 2 V and 100 μN respectively obtained by using the optimized analytical design parameters. The optimized design parameters are also used to simulate the FEM modelling of single beam RF MEMS switch. FEM models of the single beam RF MEMS switches are developed by using the optimized design parameters and presented to verify the analytical model. The results of the theoretical and numerical models are found to be in very good agreement.

In further investigations, bridge type RF MEMS switch need to design for analysis the nonlinear contact behaviours. For reliable operation of the RF MEMS switches, the FEM contact model need to investigate and contact part need to be optimized to maintain desirable contact force. This task will to be highly challenging and remarkable work in RF communication.

6. References

[1] Gabriel M. Rebeiz 2003 *RF MEMS Theory, Design and Technology*, John Wiley sons,Inc., first Edition.

[2] Dong-Ming Fang, Xiu-Han Li, Quan Yuan and Hai-Xia Zhang October 2010 Design Simulation, and Characterization of Variable Inductor with Electrostatic Actuation Fabricated by Using Surface Micromachining Technology, *IEEE transactions on electron devices*, Vol. **57**:p. 25–26.

[3] Nipun Sinha Rashed Mahameed, Marcelo B Pisani and Gianluca Piazza September 2008 Dual beam actuation of piezoelectric AlN RF MEMS switches monolithically integrated with AlN contour-mode resonators, *Journal of micromechanics and microengineering*.

[4] Montserrat Fernandez-Bolanos Badia, Elizabeth Buitrago and Adrian Mihai Ionescu 2012 RF MEMS Shunt Capacitive Switches Using AlN Compared to Si_3N_4 Dielectric, *journal of microelectromechanical systems,* **Vol. 21**:p. 273–285.

[5] Ronald G.Polcawich 2007 *Design, fabrication, test and evaluation of RF MEMS series switches using lead zirconate titanate (PZT) thin film actuators*, **Dessertation** , The Pennsylvania State University, Department of Electronic and Information Technology.

[6] Ehab K.I.Hamad, Galal E.Nadim, and Abbas S.Omar April 2004, A Proposed π-Structure RF MEMS Switch for Wide Bandwidth and High Isolation Applications", *Applied Computational Electromagnetics*.

[7] Joachim Oberhammer and Goran Stemme January 2004 Low-Voltage High-Isolation DC-to-RF MEMS Switch Based on an S-shaped Film Actuator, *Applied Computational Electromagnetics*.

[8] S.J.Gross, Q.Q.Zhang, S.Tadigadapa, S.Trolier McKinstry, T.N.Jackson and F.Djuth 2001 Reliable integration of piezoelectric lead zirconate titanate With MEMS fabrication processes, *Proc. Of SPIE*, Vol. **4558**: p. 72–80.

[9] X.Chen, C.H.J.Fox and S.Mcwilliam November 2004 Optimization of a Cantilever Microswitch with Piezoelectric Actuation, Journal of *intelligent material Systems and structures*, Vol. **15**:p. 25–26.

[10] Hee-Chul Lee, Jae-Hyoung Park, Jae-Yeong Park, Hyo-Jin Namand Jong uk Bu August 2005 Design, fabrication and RF performances of two different types of piezoelectrically actuated Ohmic MEMS switches, *Journal of Micromechanics and Microengineering*, Vol. **15** (11).

[11] J.Mehner 1993 *Mechanische Beanspruchungsanalyse von Siliziumsensoren und aktoren unter dem Einfluss von elektrostatischen und Temperatur-feldern*, **Dessertation**, TU Chemnitz.

Acknowledgments

This research was supported by the Estonian Research Council through the Institutional Research Projects IUT19-11 and PUT1435, and by the Horizon 2020 ERA-chair Grant "Cognitive Electronics COEL" —H2020-WIDESPREAD-2014-2 (Agreement 668995; TTU VFP15051).

IOP Conf. Series: Materials Science and Engineering **362** (2018) 012003 doi:10.1088/1757-899X/362/1/012003

Effect of Homo-buffer Layers on the Properties of Sputtering Deposited Ga_2O_3 Films

Jian Huang, Bing Li, Yuncheng Ma, Ke Tang [1], Haofei Huang, Yan Hu, Tianyu Zou and Linjun Wang [8]

School of Materials Science and Engineering, Shanghai University, No. 333 Nanchen Road, Shanghai, 200444, China
1.tangke@shu.edu.cn (corresponding author)

Abstract. β- Ga_2O_3 films were grown by radio-frequency magnetron sputtering method. The influence of Ga_2O_3 buffer layers and annealing treatment on the structural, optical, morphological and electrical properties of Ga_2O_3 films was studied. The results revealed an improvement of crystalline quality and transmittance of annealed β- Ga_2O_3 films prepared with homo-buffer layers. Ga_2O_3 film UV photodetectors were fabricated with a new B and Ga co-doped ZnO films (BGZO)/Au interdigitated electrode. A good ohmic contact was formed between the film and the electrode. For the detector based on Ga_2O_3 films with buffer layers, a higher value of photo response and faster response times was obtained.

1. Introduction

Ga_2O_3 has five different crystalline phases depending on ambient conditions, including α(corundum-hexagonal)-, β(monoclinic)-, γ(cubic)-, δ(cubic)-, and ε(orthorhombic)- Ga_2O_3. Among them, monoclinic structured β- Ga_2O_3 is the most stable form [1]. β- Ga_2O_3 presents outstanding properties such as wide band gap (~4.9 eV), high mechanical strength, high breakdown field strength, high chemical and thermal stability [2, 3]. Therefore, β- Ga_2O_3 has recently attracted much attention for its potential applications in optoelectronic devices [4], gas sensors [5], field effect transistors [6], transparent conductors [7] and solar-blind photodetectors [8].

As is known to all, the performance of the optoelectronic device is closely related to the film crystalline quality [9]. However, high quality β-Ga_2O_3 films were hard to be prepared by depositing directly on commonly used substrates such as glass, sapphire, Al_2O_3 and Si because of the lattice mismatch and different coefficient of thermal expansion [2, 10]. In order to improve the films crystalline quality, the introducing of a buffer layer (hetero-buffer or homo-buffer layer) to decrease the lattice mismatch between films and substrates is a well-known method [11-12].

Up to now, there are no reports on the preparation of Ga_2O_3 films with homo-buffer (*HB*) layers. In this paper, attempts were made to utilize Ga_2O_3 *HB* layers to obtain high quality Ga_2O_3 films by using radio-frequency (RF) magnetron sputtering mehtod. The influence of *HB* layers and annealing treatment on the performances of Ga_2O_3 films and Ga_2O_3 film solar-blind UV photodetectors were investigated in detail.

2. Experimental

Ga_2O_3 films were deposited by RF magnetron sputtering with Ga_2O_3 ceramic targets (purity of 99.99%). Quartz glasses with area of 20×20 mm^2 and the thickness of 1.5mm were used to be the substrate for depositing the Ga_2O_3 film. First of all, cleaning the quartz glasses with acetone and alcohol about 15 min by the ultrasonic then being dried by N_2. Before the films preparation, the vacuum chamber was evacuated until the residual gas pressure less than 5×10^{-7} Torr. For all the films deposition, the sputtering gas was pure argon (Ar) and the sputtering pressure was 6mTorr.

The growth of Ga_2O_3 films were prepared by a two-step process. Firstly, the Ga_2O_3 *HB* layer with a thickness of 30 nm was prepared on quartz glasses substrates at low temperature of 200 °C and low sputtering power of 80W. Then, Ga_2O_3 films were deposited with a thickness of 300nm at high temperature of 500 °C and high sputtering power of 150W on the *HB* layers. The films were annealed at 1000 °C for 60 min in nitrogen. Ga_2O_3 films solar-blind UV photodetectors were fabricated with new B and Ga co-doped ZnO films (BGZO)/Au (50nm / 100nm) interdigitated electrodes deposited by RF magnetron sputtering and e-beam evaporation, respectively.

The structure of the Ga_2O_3 films was characterized by X-ray diffraction (XRD, D/MAX-2200V PC, CuKα1). The surface morphology of the Ga_2O_3 films was measured by scanning electron microscope (SEM, Apollo 300). The optical properties of the films were investigated by using a UV–visible spectrophotometer (Shimadzu UV-2501PC). The photoelectric properties of the Ga_2O_3 UV detectors were measured by Keithley 2400 and a PTI optical system.

3. Results and Discussion

Figure.1 shows XRD patterns of as-grown and annealed Ga_2O_3 films prepared with and without *HB* layers. For the Ga_2O_3 film directly deposited on substrates without *HB* layer, no distinct peaks were observed with the exception of a broad peak originating from the substrates, indicating the formation of

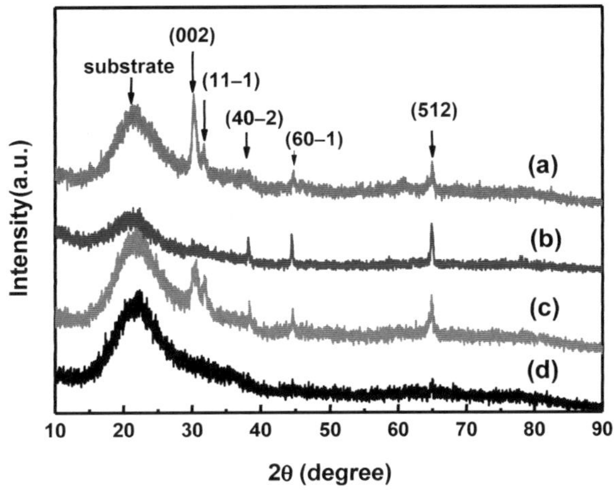

Figure 1. XRD patterns of Ga2O3 films deposited in different conditions: (a) annealed with HB layers, (b) as-grown with HB layers, (c) annealed without HB layers, (d) as-grown without HB layers

amorphous Ga_2O_3 film. After annealing treatment, five peaks located at 31.94°, 33.08°, 37.82 °, 44.08° and 64.48° corresponding to monoclinic structured β-Ga_2O_3 (002), $(11\bar{1})$, $(40\bar{2})$, $(60\bar{1})$ and (512) appear (JCPDS No. 43-1012), indicating the initial amorphous structure evolved into the polycrystalline structure [13]. For as-grown Ga_2O_3 films with *HB* layers, three peaks corresponding to β-Ga_2O_3 $(40\bar{2})$, $(60\bar{1})$ and (512) are observed. For all the prepared Ga_2O_3 films, annealed Ga_2O_3 films with *HB* layers show the stronger diffraction intensity of (002) peak. The results indicate that thc *HB* layer and annealing treatment can improve the polycrystalline quality of Ga_2O_3 films significantly.

Figure.2 shows SEM images of as-grown and annealed Ga_2O_3 films prepared with and without *HB* layers. The as-grown Ga_2O_3 film without *HB* layers shows a uniform and dense surface with small grain size. After annealing treatment, the grain size of Ga_2O_3 films grows slightly larger. It can be obviously found, compared with samples without *HB* layers, the Ga_2O_3 films grown on *HB* layers show larger grain size, indicating the improvement of film crystalline quality which is consistent with the XRD results.

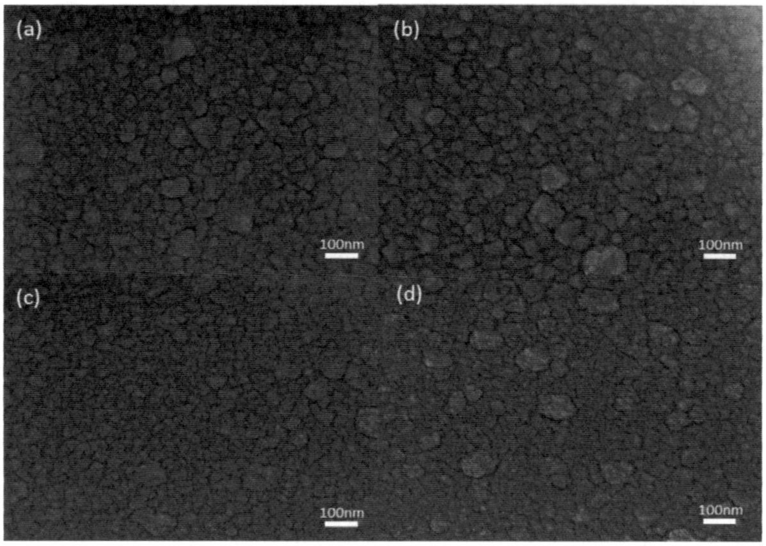

Figure 2. SEM images of Ga_2O_3 films: (a) annealed without *HB* layers, (b) annealed with *HB* layers, (c) as-grown without *HB* layers and (d) as-grown with *HB* layers

The transmittance spectra and absorption spectra in the wavelength range 200-800 nm for annealed Ga_2O_3 films with and without *HB* layers are shown in Figure. 3 (a). Both samples (including the substrates) show high average transmittance over 85%. It can also be found that the transmittance of Ga_2O_3 films with *HB* layers is slightly higher than those without *HB* layers, which may be due to the better crystallinity and good uniformity of the films with buffer layers. From the absorption spectra, the fundamental absorption edges of the films can be found clearly. Figure.3(b) shows the plots of $(\alpha h\nu)^2$ versus $h\nu$ for annealed Ga_2O_3 films with and without *HB* layers, where α is the optical absorption coefficient and $h\nu$ is photon energy. As a direct band gap semiconductor, the optical energy band gap (E_g) values of the Ga_2O_3 films can be calculated using Tauc's formula[14]:

$$\alpha h\nu = C(h\nu - E_g)^{1/2} \tag{1}$$

Where C is a constant. The E_g can be obtained by extrapolating the linear region of the plot $(\alpha h\nu)^2$ versus $h\nu$ and taking the intercept on the $h\nu$-axis [14]. And it is obvious shown in Figure.3(b) that the optical energy band gap of films with and without homo-buffer layers is 4.95eV and 4.87eV, respectively.

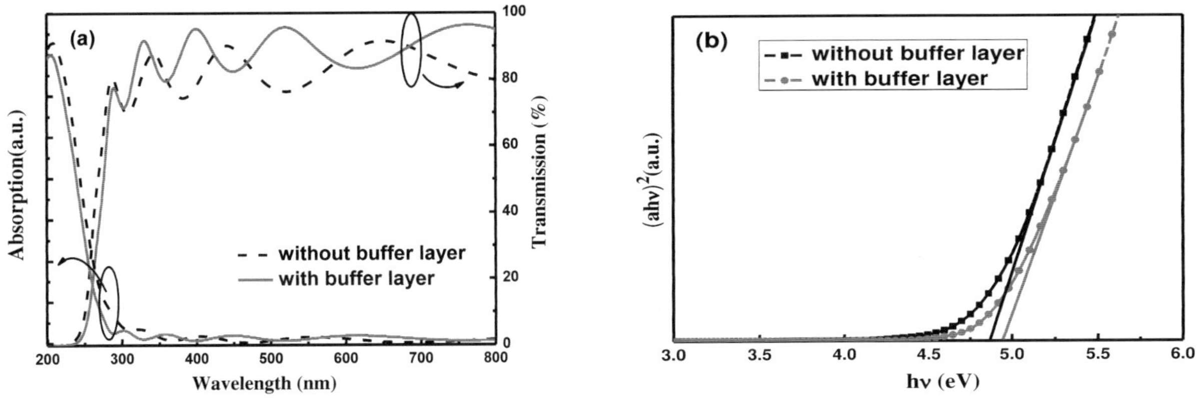

Figure 3. The transmission spectra and the absorption spectra(a) and the (αhv)2 vs. hv plots for Ga2O3 films (b)

Figure.4 shows the room temperature *I-V* characteristics of Ga_2O_3 films detectors with and without homo-buffer layers in dark. To evaluate the contact characteristics, the *I-V* plots were fitted using non-linear function: $I(V) = aV^b$, where a is constant, b is an ohmic resistance coefficient. If b=1, the contact is ohmic [15]. The calculated value b is about 1.0010 which is much close to 1. The results confirm the ohmic contacts between BGZO/Au and Ga_2O_3 films. It can also be found clearly that the dark current of detectors with homo-buffer layer reduce 3 times due to the improved crystalline quality with less grain boundaries and defects.

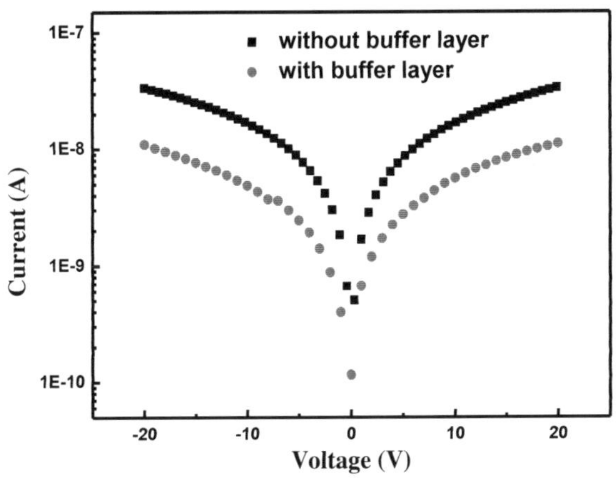

Figure 4. logarithmic *I-V* properties of β- Ga_2O_3 films detectors in dark

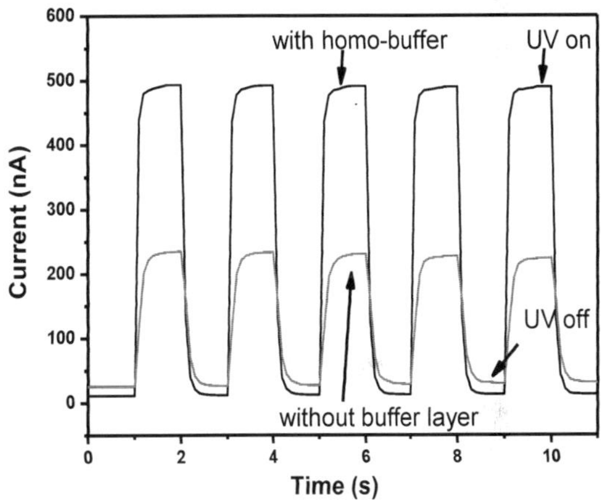

Figure 5. Time-dependent photoresponse of the β- Ga_2O_3 detector to UV light illumination

Figure.5 shows time-dependent response of the detectors with and without *HB* layers measured under bias of 20 V with turning on and off as 250 nm light irradiated on the device. Upon UV illumination, the photocurrent for detectors with and without *HB* layers increased to about 490 and 230nA, respectively. As turn the UV light off, it decreased dramatically to the initial value about 12nA and 30nA, respectively. The detectors exhibit a nearly identical response after multiple illumination cycles indicating that the device has good reproducible characteristics. For Ga_2O_3 detectors with and without *HB* layers, the UV light to dark current ratio is about 41 and 8, respectively.

The energy of the UV (250nm) photons (~5eV) is larger than the optical energy band gap of Ga_2O_3 films. When UV light illuminates the detectors, the UV photons are absorbed in Ga_2O_3 films, generating electron-hole pairs, responsible for the photocurrent in detectors. In the sputtered polycrystalline Ga_2O_3 films, there are inevitably many defects and grain boundaries acting as the trapping centers which is also named carrier trapping effect [16]. Before being separated and collected by electrodes, the generated carriers in Ga_2O_3 film detectors could be seized by those trapping centers. Therefore, the film crystalline quality strongly affects the properties of detectors, which is the reason for difference in UV light to dark current ratio of Ga_2O_3 UV detectors. The better crystalline quality of Ga_2O_3 films with *HB* layers resulting in the higher UV light to dark current ratio of detectors. Rise time (t_R) and fall time (t_F) is defined as the time taken to vary between 10% and 90% of the maximum photocurrent. From Figure. 5, t_R is about 0.1s and 0.15s, and t_F is about 0.2s and 0.3s for detectors with and without *HB* layers respectively. The faster response times for detectors with *HB* layers are mainly attributed to the higher crystalline quality of Ga_2O_3 films.

4. Summary

A two-step growth regime was used to achieve an epitaxial growth of β- Ga_2O_3 films by RF magnetron sputtering method. The influence of homo-buffer layers and annealing treatment on the properties of Ga_2O_3 films was investigated. The results indicate that the homo-buffer layers and annealing treatment can improve the polycrystalline quality of Ga_2O_3 films. The films with homo-buffer layers show larger grain size and higher transparency. The optical band gap of films with homo-buffer layers is about 4.95eV. The higher value of UV light to dark current ratio and better time-dependent characteristic of the Ga_2O_3 detectors with homo-buffer layers manifest that the film crystalline quality plays an important role on the performance of detectors.

5. References

[1] Feng X J, Li Z, Mi W, Luo Y and Ma J 2015 Mg-doped β-Ga_2O_3 films with tunable optical bandgap prepared on MgO(110) substrates by metal-organic chemical vapor deposition, *Materials Science in Semiconductor Processing* **34** 52-57.

[2] Choi K H and Kang H C 2014 Structural and optical evolution of Ga_2O_3/glass thin films deposited by radio frequency magnetron sputtering, *Materials Letters* **123** 160-164.

[3] Kokubun Y, Miura K, Endo F and Nakagomi S 2007 Sol-gel prepared β- Ga_2O_3 thin films for ultraviolet photodetectors, *Appl. Phys. Lett.* **90** 031912.

[4] Guo X C, Hao N H, Guo D Y, Wu Z P, An Y H, Chu X L, Li L H, Li P G, Lei M and Tang W H 2016 β- Ga_2O_3/p-Si heterojunction solar-blind ultraviolet photodetector with enhanced photoelectric responsivity, *Journal of Alloys and Compounds* **660** 136-140.

[5] Li Y, Trinchi A, Wlodarski W, Galatsis K and Kalantar-zadeh K 2003, Investigation of the oxygen gas sensing performance of Ga_2O_3 thin films with different dopants, *Sens. Actuators B Chem.* **93** 431-434.

[6] Higashiwaki M, Sasaki K, Kuramata A, Masui T and Yamakoshi S 2012,Gallium oxide (Ga_2O_3) metal-semiconductor field-effect transistors on single-crystal β-Ga_2O_3(010) substrates, *Appl. Phys. Lett.* **100** 013504.

[7] Ueda N, Hosono H, Waseda R, and Kawazoe H 1997 Synthesis and control of conductivity of ultraviolet transmitting β- Ga_2O_3 single crystals, *Appl. Phys. Lett.* **70** 3561.

[8] Rikiya S, Shinji N and Yoshihiro K 2011 Solar-blind photodiodes composed of a Au Schottky contact and a Ga_2O_3 single crystal with a high resistivity cap layer, *Appl. Phys. Lett.* **98** 131114.

[9] Zyss J 2013 Molecular nonlinear optics: materials, physics, and devices[M]. Academic press.

[10] Feng X J , Li Z, Mi W and Ma J 2016 Effect of annealing on the properties of Ga_2O_3: Mg films prepared on α-Al_2O_3 (0001) by MOCVD. *Vacuum* **124** 101-107.

[11] Ting S Y, Chen P J, Wang H C, Liao C H, Chang W M, Hsieh Y P, Yang C C 2012 Crystallinity improvement of ZnO thin film on different buffer layers grown by MBE. *Journal of Nanomaterials* 2817-2827.

[12] Liu M, Bi D, Man B Y,Kong D M, Xu X Y 2013 Effects of CdZnTe buffer layer thickness on properties of HgCdTe thin film grown by pulsed laser deposition. *Applied Surface Science* 264 522-526.

[13] Guo D Y, Wu Z P, Li P G, An Y H, Liu H, Guo X C, Yan H,Wang G F, Sun C L, Li L H, Tang W H 2014 Fabrication of β- Ga_2O_3 thin films and solar-blind photodetectors by laser MBE technology.*Optical Materials Express* **4** 1067-1076.

[14] Igasaki Y and Saito H 1991 Substrate temperature dependence of electrical properties of ZnO: Al epitaxial films on sapphire (1210). *Journal of Applied Physics* **69** 2190-2195.
[15] Choi B W and Wadley H N G 2000 In situ studies of Cd1-xZnxTe nucleation and crystal growth. *Journal of Crystal Growth* **208** 219-230.
[16] Souw E K and Meilunas R J 1997 Response of CVD diamond detectors to alpha radiation, *Nuclear Instruments and Methods in Physics Research Section A* **400** 69-86.

Acknowledgements
This work was funded by Technology Commission of Shanghai (No. 16010500500).

IOP Conf. Series: Materials Science and Engineering **362** (2018) 012004 doi:10.1088/1757-899X/362/1/012004

Properties of ZnTe Films Deposited with Different Substrate Temperature Using CSS Method

Qingmiao Gu, Jian Huang[1]*, Yuncheng Ma, Ke Tang[2]*, Haofei Huang, Yan Hu, Tianyu Zou and Linjun Wang

School of Materials Science and Engineering, Shanghai University, No. 333 Nanchen Road, Shanghai, 200444, China
E-mail:[1*]jianhuang@shu.edu.cn(corresponding author),
[2*]dtangke@shu.edu.cn(corresponding author)

Abstract. ZnTe films were prepared by close spaced-sublimation (CSS) method. The effects of substrate temperature on ZnTe films were investigated. The results showed cubic structure with strong (111) texture for all the films. Films with higher crystalline quality and larger grain size were obtained at higher substrate temperature. A red-shift of the optical bandgap and reduction of resistivity were observed with increasing substrate temperature.

1. Introduction

Zinc telluride (ZnTe), an important II-VI semiconductor, has recently received an increasing attention due to its excellent properties such as direct wide band-gap of 2.26 eV at room temperature, low electron affinity of 3.53 eV, high absorption coefficient of 105 cm−1 and high electro-optic coefficient, which makes it a potential candidate for applications in optoelectronic devices and integrated optics such as solar cell, light emitting diodes, terahertz detectors [1-4].

To get Device-grade of ZnTe, a variety of film deposition techniques have to be applied [5], including thermal evaporation [6, 7], RF and DC sputtering [8], close spaced-sublimation (CSS) [9] and etc. Among those, CSS is a promising preparation technique for film deposition in terms of its rapid deposition rate and simple configuration. CSS method also offers many possibilities to adjust the preparation conditions to obtain uniform surface and large area thick films [10]. Up to now, relatively few results have been reported on preparation of ZnTe films using CSS method. The previous studies show that the substrate temperature for many deposition techniques is one of the key deposition parameters which has significant effects on the properties of ZnTe film [11].

In this paper, ZnTe films were deposited on glasses by CSS method. Substrate temperature dependent on properties of the ZnTe films were investigated systematically.

2. Experimental

ZnTe films were deposited on quartz glasses for 30 min with different substrate temperatures by CSS method. The sublimation source material was high-purity ZnTe powder (99.99%). The chamber was pumped down to the vacuum of 5×10^{-3} Pa before deposition. During the deposition, the source temperature was kept at 650 ^{o}C and the spacing between source and substrate was about 3 mm. The substrate temperature was varied from 200 ^{o}C to 500°C.

The crystallinity of the films was evaluated by X-ray diffraction (XRD, D/MAX-2200, Cu Kα, λ=0.15406 nm). The surface morphology was characterized by field-emission gun scanning electron microscopy (FESEM, Apollo 300) and the ZnTe composition was estimated using energy dispersive

IOP Conf. Series: Materials Science and Engineering **362** (2018) 012004 doi:10.1088/1757-899X/362/1/012004

spectrometer (EDS). Au electrode dots with a diameter of 2 mm were prepared by double ion beam sputtering, and current density-voltage (J-V) curves were measured using Keithley 2400 source meter. The absorption and transmission of the ZnTe films were measured by Jasco UV-570 spectrophotometer.

3. Results and Discussion

Figure1 shows the XRD patterns of the ZnTe films. Compared with standard PDF card No. 15-0746, all ZnTe films exhibit a zinc blende structure with strong preferential (111) orientation. Most notably, the relative intensity of (111) peak increases apparently with increasing Ts. It suggests the better crystallinity and preferential orientation at higher Ts. Figure 2 shows a function of Ts with the position and full width at half maximum (FWHM) of the (111) peak. As shown in the Figure 2, the (111) peak positions are located at 2θ=25.019°, 25.280°, 25.259°, 25.221° for the films deposited at different substrate temperature, respectively. Compared with the case of Ts =200°C, the peak position of ZnTe (111) diffraction obviously shifts to large angle when Ts above 200 °C, which is very close to the standard value (2θ=25.259°) mentioned in JCPDS (15-0746). The figure also shows that the FWHM of (111) peaks decreases with increasing Ts, which implies the higher crystalline quality of the films at higher Ts.

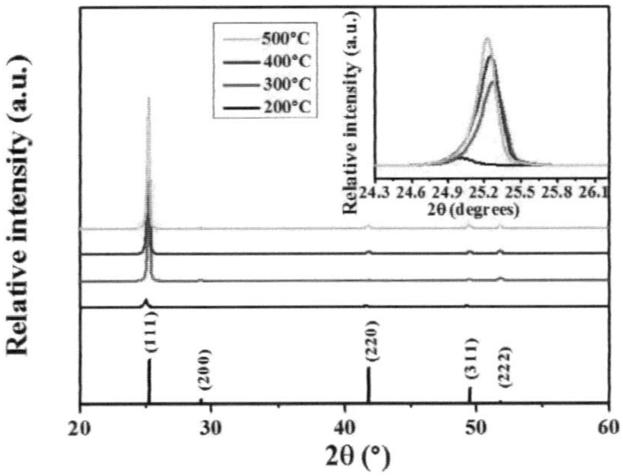

Figure 1. XRD patterns of the ZnTe films deposited at different substrate temperature (Ts). The inset is the enlargement of the (111) peak.

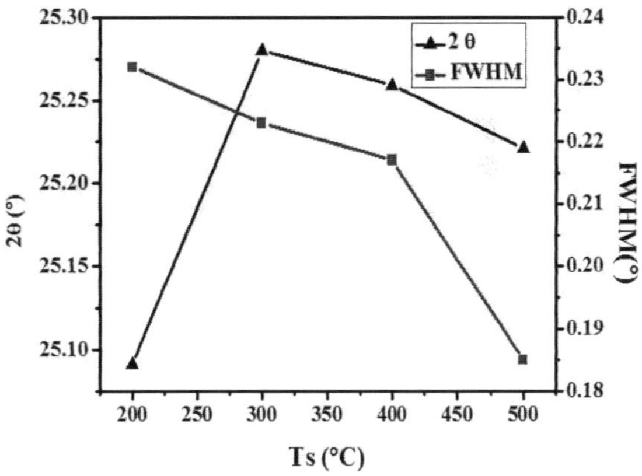

Figure 2. A function of Ts with the position and FWHM of the (111) peak.

The variation in grain size and FWHM of the peak corresponding to the (111) plane as a function of T_s is listed in Table 1. The grain size is calculated from the Scherrer relation [12]:

$$D = \frac{0.89\lambda}{\beta COS(\theta)} \tag{1}$$

where D is the grain size of ZnTe films, β is the FWHM of the (111) peak, λ is the wavelength of

Table 1. The FWHM and grain size of the ZnTe films at different substrate temperature (T_s).

T_s [°C]	2θ [°]	FWHM [°]	grain size(D) [nm]
200	25.019	0.232	34.672
300	25.280	0.223	36.090
400	25.259	0.217	37.088

500	25.221	0.185	43.4944

the X-rays (CuKα1 1.54056Å), and θ is the diffractive angle. From the Table 1, we can find that the grain size increases with increasing Ts, which indicates that the higher T_s can get the better crystallinity.

Figure 3 shows the SEM images of ZnTe films deposited at different T_s From Figure 3, we can find that, with increasing T_s the grain size increases, which is consistent with the XRD results.

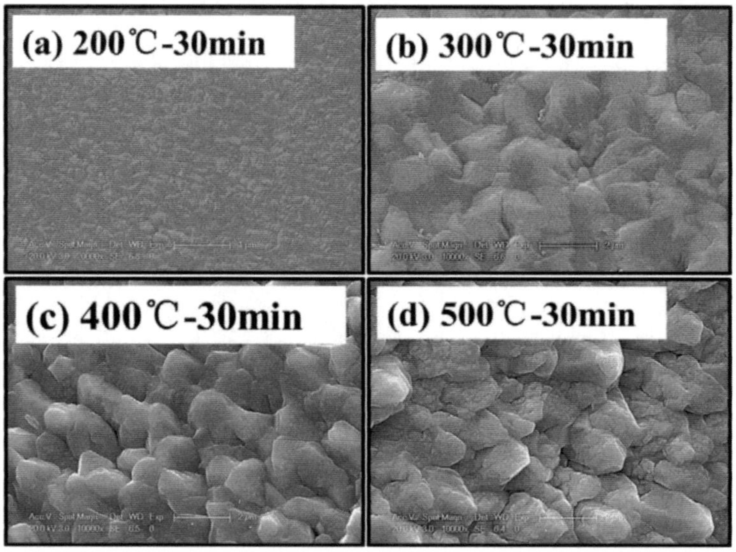

Figure 3. SEM images of the ZnTe films obtained at different substrate temperature.

It is well known that electrical properties of ZnTe films strongly depend on composition stoichiometry [13]. Table 2 presents the composition of ZnTe films obtained under different substrate

Table 2. Composition and atomic ratio for ZnTe films obtained from different conditions.

Samples	***Ts*** **[°C]**	**Atom rate [%]**		**Zn: Te**
		Zn	**Te**	
Deposited-Sample	200	54.76	45.24	1.21
	300	48.79	51.21	0.95
	400	49.32	50.68	0.97
	500	49.16	50.84	0.97

temperature estimated using energy dispersive spectrometer (EDS). We can find at T_s =200 °C, the atomic ratio (Zn:Te=1.21) is greater than 1. However, above 200 °C, the ZnTe atomic ratio is close to the 1. It is proposed that low substrate temperatures is a highly non-equilibrium conditions, which might produce defects in II-VI films [14]. The result indicates the improved crystalline quality of the ZnTe films at high T_s.

The absorption spectra in the wavelength range of 450–700 nm for the ZnTe films obtained from different conditions are given in Figure 4. The fundamental absorption edges are clearly observed for

IOP Conf. Series: Materials Science and Engineering **362** (2018) 012004 doi:10.1088/1757-899X/362/1/012004

all samples. The absorption edge of the ZnTe films shifts to the longer wavelength with T_s increasing from 200 °C to 500 °C.

The optical absorption at absorption edge corresponds to the electron transition from valence band to conduction band. The optical band-gap (E_g) of materials can be estimated from the change of the absorption coefficient (α) in this edge region by using the function expressed as follows [15]:

$$\alpha h\upsilon = B(h\upsilon - E_g)^n \quad (2)$$

where h is Plank's constant, v is the frequency of the radiation, B is a constant which depends on the nature of transition and n is a number which can take the values 1/2, 3/2, 2 or more depending on

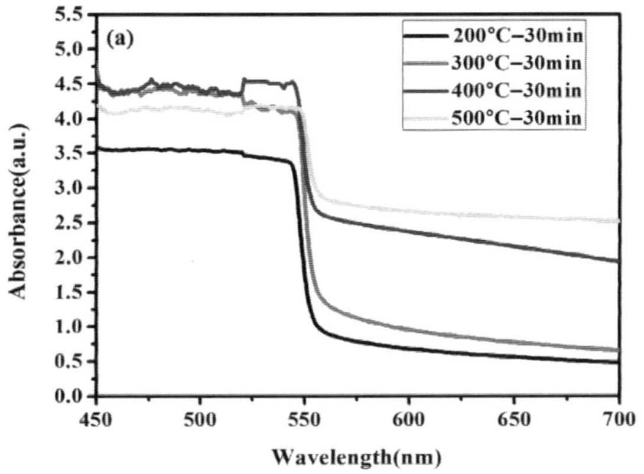

Figure 4. The absorption spectra of the films obtained from different substrate temperature.

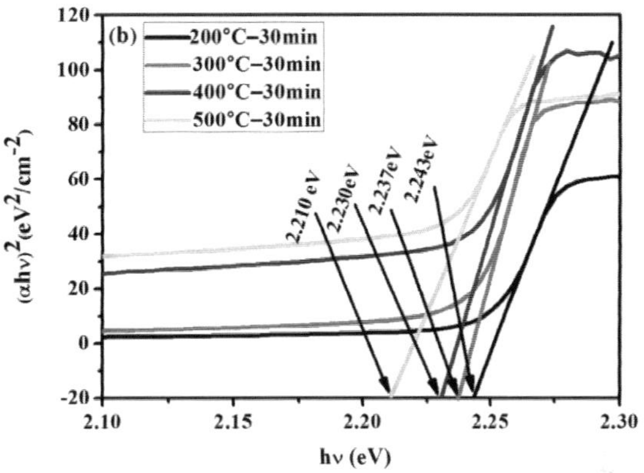

Figure 5. The $(\alpha h v)^2$ *vs.* hv plots for films obtained from different substrate temperature.

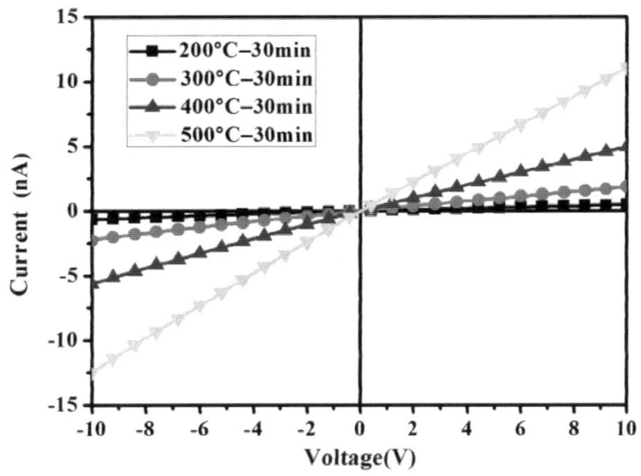

Figure 6. The *I-V* curves for films obtained from different conditions.

whether the transition is direct-allowed, direct-forbidden, indirect-allowed or indirect-forbidden. By extrapolating the linear portion of the curves, we can get the optical band-gap, as shown in the Figure 5. It can be found that, with increasing T_s, the film optical band-gap decreases.

The I-V plots of the films deposited at different substrate temperatures was shown in Figure 6. A linear relationship can be observed indicating ohmic contacts between ZnTe films and Au electrodes. Compared the I-V curves of different samples, we can observe that the electrical resistivity of the films decreases with substrate temperature.

4. Summary

ZnTe films were deposited on to cleaned quartz glasses by CSS with different substrate temperature (200-500 °C). XRD results shows that ZnTe films obtained have zinc-bende structure, and high T_s are easier to obtain high-quality films. Films with higher crystalline quality and larger grain size were obtained at higher substrate temperature. All results indicate that increasing the substrate temperature can improve the quality of the ZnTe films effectively.

5. References

[1] Nishio M, Hayashida K, Guo Q and Ogawa H 2001 Growth rate characteristics and photoluminescence properties of ZnTe in MOVPE system, *Appl. Sur. Sci.* **169** 227-230.

[2] Arya S, Khan S, Lehana P, Gupta I and Kumar S 2014 Electrical properties of electrodeposited zinc selenide (ZnSe) nanowires, *J. Mater. Sci: Mater. Elec.* **25** 4150-55.

[3] Sato K and Adachi S 1993 Optical properties of ZnTe, *J. Appl. Phys.* **73** 926-931.

[4] Rioux D, Niles D W and Höchst H 1993 ZnTe: A potential interlayer to form low resistance back contacts in CdS/CdTe solar cells, *J. Appl. Phys.* **73** 8381-85.

[5] Rao G K, Bangera K V and Shivakumar G 2009 The effect of substrate temperature on the structural, optical and electrical properties of vacuum deposited ZnTe thin films, *Vacuum* **83** 1485-88.

[6] Nauer M, Ernst K, Kautek W and Neumann-Spallart M 2005 Depth profile characterization of electrodeposited multi-thin-film structures by low angle of incidence X-ray diffractometry, *Thin Sol. Films* **489** 86-93.

[7] Aqili A K, Ali Z and Maqsood A 2000 Optical and structural properties of two-sourced evaporated ZnTe thin films, *Appl. Sur. Sci.* **167** 1-11.

[8] Compaan A D, Gupta A, Lee S, Wang S and Drayton J 2004 High efficiency, magnetron sputtered CdS/CdTe solar cells, *Solar Energy* **77** 815-822.

[9] Tokuda S, Kishihara H, Adachi S and Sato T 2004 Preparation and characterization of polycrystalline CdZnTe films for large-area, high-sensitivity X-ray detectors, *J. Mater. Sci: Mater. Elec.* **15** 1-8.

[10] Bacaksiz E, Basol B, Altunbaş M, Novruzov V, Yanmaz E and Nezir S 2007 Effects of substrate temperature and post-deposition anneal on properties of evaporated cadmium telluride films, *Thin Sol. Films* **515** 3079-84.

[11] Rao G K, Bangera K V and Shivakumar G K 2009 The effect of substrate temperature on the structural, optical and electrical properties of vacuum deposited ZnTe thin films, *Vacuum* **83** 1485-88.

[12] Catharine H M and Speller R D 2000 A CdZnTe array for the detection of explosives in baggage by energy-dispersive X-ray diffraction signatures at multiple scatter angles, *Nucl. Inst. and Meth. Phys. Res. Sec. A: Acce., Spec., Dete. Asso. Equi.* **449** 408-415.

[13] Belyaev A P, Rubets V P, Antipov V V and Grishin V V 2008 Electrical and galvanomagnetic properties of cadmium telluride films synthesized under highly nonequilibrium conditions, *Semi.* **42** 1282-85.

[14] Chopra K L and Das S R 1983 Why Thin Film Solar Cells, *Thin Films Solar Cells* 1-18.

[15] Razykov T M, Karazhanov S Z, Leiderman A Y, Khusainova N F and Kouchkarov K 2006 Effect of the grain boundaries on the conductivity and current transport in II–VI films, *Sol. Ener. Mater. & Sol. Cel.* **90** 2255–62.

Acknowledgements
This work was funded by Science and Technology Commission of Shanghai (No. 16010500500, 15520500200).

IOP Conf. Series: Materials Science and Engineering **362** (2018) 012005 doi:10.1088/1757-899X/362/1/012005

Preparation and Dynamic Mechanical Analysis of Glass or carbon Fiber/Polymer Composites

Konstantinos Karvanis*, Soňa Rusnáková, Milan Žaludek, AlexanderČapka

Department of Production Engineering, Faculty of Technology
Tomas Bata University in Zlín
Vavrečkova 275, 760 01, Zlín, Czech Republic

*e-mail: karvanis@utb.cz

Abstract. Nowadays, glass fiber reinforced polymer (GFRP) and carbon fiber reinforced polymer (CFRP) composites are broadly used so it is essential their properties to be explored in depth. In this study, GFRPs and CFRPs were produced by vacuum bag oven method and their viscoelastic behaviour at elevated temperatures was investigated. In particular, by Dynamic Mechanical Analysis experiments properties such as storage modulus, loss modulus, tanδ and glass transition temperature were determined. The results showed that, apart from a very small increment in the low temperatures in some of the compounds, as the temperature increases the storage modulus of the composites decrease while the composite containing unidirectional glass fibers in longitudinal direction and 34% percentage of matrix achieved the higher storage and loss modulus. Furthermore, as the unidirectional fibers change direction from transverse to longitudinal the GFRPs exhibit much higher storage and loss modulus.

1. Introduction

Composites are being used by humans from ancient times. Their great advantage is that they combine the properties of their individual substances so the mechanical behaviour of the final composite is better than the one of the initial materials. A great category of them is the fiber reinforced polymer (FRP) composites which are composed of fibers, such as glass or carbon, embedded in a polymer matrix. Types of them are the glass fiber reinforced polymers (GFRPs) and carbon fiber reinforced polymers (CFRPs) while both of them are familiar for their low weight, high strength and modulus. The applications of these compounds involve a broad range such as in automotive, aircraft, marine and space industry.

Furthermore, prepregs materials can be used in order to achieve maximum performance of the composites. Prepregs are materials which have been pre impregnated with the resin and they are ready to be cured under heat and pressure combination. Advantages such as excellent quality of the final product, high fibers percentage, determination of the exact ratio of matrix-reinforcement percentage and many other have done the prepregs very popular in the composites industry while their high cost is a restrictive factor for their extensive use.

Generally, the production of the composites by prepregs can be separated in autoclave and out of autoclave (OOA) methods. By autoclave use, high quality compounds are produced. It should be noted that, especially for big parts, autoclaves imply considerable costs for function and tooling [1].

Nowadays, scientists are trying to develop new techniques such as the vacuum bag oven process, without the increased cost of the autoclave's use. During the vacuum bag oven technique, the composite is placed in a mould under a vacuum bag, the air is evacuated from the system and the cure of the compound takes place in an oven. An important factor which must be taken into account in OOA methods is to avoid the porosity and form of voids in the final composite due to the present of air during its production.
As a result of the broadly use of GFRPs and CFRPs, it is essential their properties to be entirely investigated so as the mechanical behaviour of them to be absolutely predictable. An important point in the investigation of the materials is how their properties are influenced by fluctuations of the temperature.
An efficient way for the exploration of the materials' mechanical behaviour is the Dynamic Mechanical Analysis (DMA) which offers study of the rheology under different temperatures, frequencies or loadings. In particular, by these experiments properties such as storage modulus, loss modulus and tanδ can be determined. Moreover, DMA offers the ability for the determination of the glass transition temperature (Tg). Below this temperature many amorphous polymer are hard and stiff while above it they are transformed to soft.
By DMA tests, researchers have investigated various properties of materials such as epoxy based glass/carbon hybrid composite laminates [2] epoxy/poly(methylmethacrylate)/ glass fiber composites [3], glass fiber reinforced polymer composites [4,5,6,7] and carbon fiber reinforced polymer composites [8,9,10]. Furthermore, Gupta and Rohit [11] explored the dynamic mechanical properties of glass fibres reinforced epoxy composites with varying numbers of layers and it was found that the highest storage modulus value was achieved by the composite with the highest number of layers. By dynamic mechanical analyser method, Kister and Dossi [12] monitored the cure characteristics of carbon/epoxy composite prepreg. Goertzen and Kessler [13] evaluated carbon fiber/epoxy matrix composite through dynamic mechanical analysis. Katunin et al. [14] studied the ageing of GFRP laminates in deionized and seawater and investigated them by various tests including DMA.
However, a broad research study on the dynamic mechanical properties of polymer matrix composites containing carbon or glass fibers in different orientations while the thickness of the compounds and the percentage of the matrix are variable seems to be missing. Specifically, it would be useful a comparison between the viscoelastic properties of GFRP composites containing unidirectional fibers in longitudinal or transverse direction.
In our study, GFRPs and CFRPs were produced and by DMA tests their dynamic mechanical properties were investigated in order to determine the optimal fibers' type and orientation under elevated temperatures. The basic scope of this study is for the GFRPs to compare unidirectional fabrics when they are placed in longitudinal or transverse direction and for the CFRPs to provide some details for their viscoelastic behaviour because these compounds have different thicknesses so the results of the tests cannot be compared directly.
Furthermore, two different methods were used for the determination of composites' Tg temperature. Particularly, the compounds were designed and tested for transport industry applications.

2. Experimental

2.1. Materials

For the production of the composites prepregs in various layers numbers were used so as the thickness of the compounds to be variable. Both carbon and glass fibers were in unidirectional or woven fabric type.
The glass fiber epoxy prepregs, both the unidirectional and woven types, were fabricated by Delta-Preg S.p.A. by the resin DT 806R. The carbon fiber prepregs were prepared by Impregnatex Compositi S.r.l. using the resin IMP 503Z. In Table I, the specifications of the prepregs and the fibers' type which were used are presented. The unidirectional were placed in longitudinal (Fig. 1) or transverse direction while the woven were oriented at 0° (fibers at $0^{o}/90^{o}$) or at 45° (fibers at $-45^{o}/45^{o}$). Specifically, the following composites were fabricated:

- GFRPs consisted of unidirectional fibers in longitudinal direction
- GFRPs consisted of unidirectional fibers in transverse direction
- GFRPs consisted of woven fabrics at 0°, (fibers 0°/90°)
- GFRPs consisted of woven fabrics at 45°,
  (fibers -45°/+45°)
- CFRPs consisted of unidirectional fibers in longitudinal direction
- CFRPs consisted of unidirectional fibers in transverse direction
- CFRPs consisted of woven fabrics 0°, (0°/90°)
- CFRPs consisted of woven fabrics 45°, (-45°/45°)

Furthermore, two more GFRPs (UD fibers in longitudinal or transverse direction) with different percentage of matrix were produced. These composites are presented analytically in Table I.

Figure 1. Base of orientation: unidirectional fibers in longitudinal direction (0°)

Table 1. Types of prepregs-fibers and specifications of composites

Symbol of composite	Type of prepreg-Fabric name	Type of fibers and orientation	Percentage of matrix	Thickness of composite (mm) +-0.1
GF-UD-L-34%	VV430U-DT806R-34	Glass fiber, unidirectional, longitudinal (0°)	34%	2.35
GF-UD-TR-34%	VV430U-DT806R-34	Glass fiber, unidirectional transverse (90°)	34%	2.35
GF-UD-L-39%	VV430U-DT806R-39	Glass fiber, unidirectional longitudinal (0°)	39%	2.85
GF-UD-TR-39%	VV430U-DT806R-39	Glass fiber, unidirectional transverse (90°)	39%	2.85
GF-W-0°-37%	VV320P-DT806R-37	Glass fiber, woven,0° (0°/90°)	37%	2.35
GF-W-45°-37%	VV320P-DT806R-37	Glass fiber, woven, 45° (-45°/+45°)	37%	2.35
CF-UD-L-38%	IMP503Z/ GV420U	Carbon fiber, unidirectional longitudinal (0°)	38%, +-1%	3.7
CF-UD-TR-38%	IMP503Z/ GV420U	Carbon fiber, unidirectional transverse (90°)	38%, +-1%	2.85
CF-W-0°-38%	IMP503Z/ GG630T	Carbon fiber, woven 0° (0°/90°)	38%, +-1%	3.15
CF-W-45°-38%	IMP503Z/ GG630T	Carbon fiber, woven 45° (-45°/+45°)	38%, +-1%	1.3

2.2. Production method

By vacuum bag oven method GFRPs and CFRPs were fabricated. The base of this process is the construction of the vacuum bag system.

Specifically, the steps described bellows were followed. Firstly, the prepreg laminates were cut in the desired dimensions and they were put, in various layers numbers, in the moulds. Over them, a release film and a breath cloth were placed. Then, the system was covered by a vacuum bag which was stuck in the perimeter of the mould by sticky tape. Also, the bag was connected with the vacuum pump by the vacuum valve. After the sealing of the system, air leak was tested as it is one of the most critical problems in these processes. Finally, the system was placed in the oven where the cure process took place by the following steps. Firstly, the temperature was increased slowly up to 65 °C where it was held for 30 minutes, subsequently the temperature was raised to 120 °C where it was maintained for 60 minutes and finally it was slowly reduced up to ambient temperature. Totally, this process lasted about 4 hours. The composites were let to solidify for one day at room temperature and then samples were mechanically cut in the desired dimensions.

Figure 2. Production of composites.

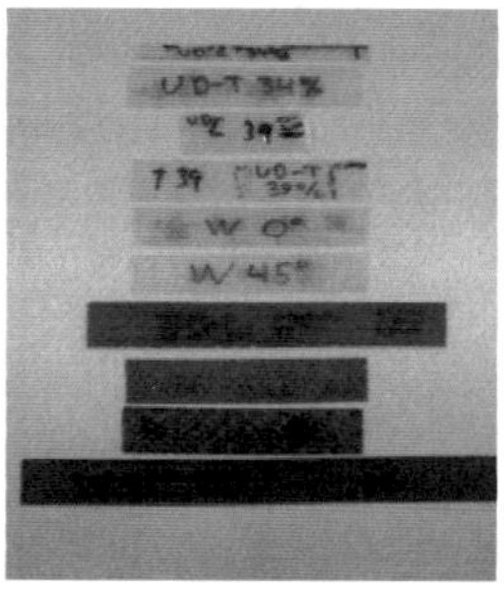

Figure 3. Plates of composites.

2.3. Experimental details

The DMA experiments were carried out by METTLER TOLEDO instrument over a temperature range of 30-200 °C, at a steady frequency of 1Hz while the heating rate was 2 °C/min. The displacement amplitude adjustment was 20 μm whereas a force of 1 N was used as preload. The samples were tested under a 3 point bending configuration mode which is suggested by METTLER TOLEDO for composite and stiff materials while the span length between the supports was 30 mm. The specimens were in rectangular shape with dimensions of (40x7) mm whereas their thickness was variable and it is presented in table I.

Figure 4. Specimens after the experiments.

3. Results

3.1. Storage modulus

The storage modulus (E') is correlative with the stored energy. Figures 5 and 6 depict the effect of the temperature on the storage modulus of GFRPs and CFRPs, respectively. As it can be seen, except for a slightly increase in the low temperatures in some of the composites, the storage modulus decreases as the temperature increases. A remarkable point is that in all of the curves there is a sharp decrease which is associated with the glass transition temperature (Tg). Usually, this steep fall determines the maximum operating temperature of the materials. Despite the fact of its relative small thickness, the GFRP compound with 34% percentage of matrix exhibited much higher storage modulus than the GFRP with 39% matrix. This can be attributed to the positive effect which the glass fibers offer to the composite and it is indicative of the fact that a small increment in the percentage of them can give a high positive change in the measured values. In all cases of GFRPs, the placement of the unidirectional fibers in the longitudinal direction seems to be much more effective than if they are placed in the transverse direction. Furthermore, the GFRP sample with woven fabric at 0° succeeded higher storage modulus than the GFRP with woven fabric at 45°.

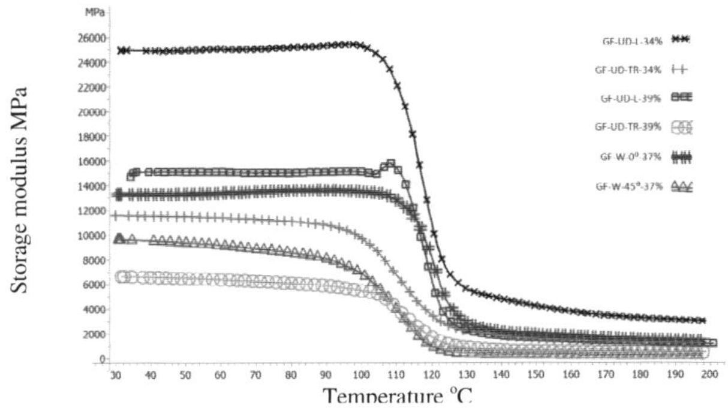

Figure 5. Storage modulus of GFRPs as a function of temperature.

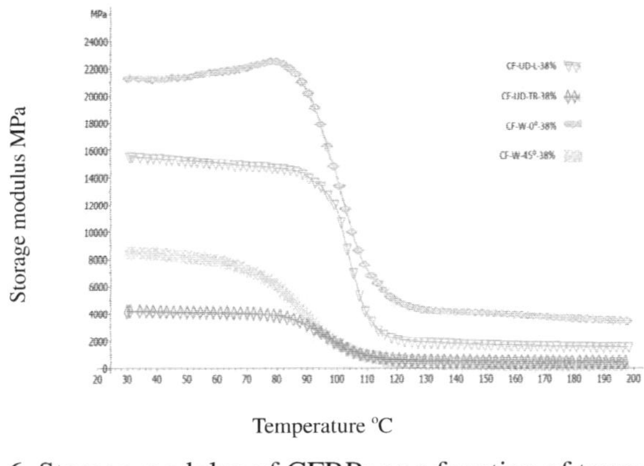

Figure 6. Storage modulus of CFRPs as a function of temperature.

3.2. Loss modulus

The loss modulus (E") is related with the energy which was dissipated as heat under dynamic mechanical analysis experiments. The values of the loss modulus at elevated temperatures of GFRPs and CFRPs are presented in figures 7 and 8. As it can be revealed from these graphs, the values are almost stable at low temperatures while after about the 70-80 °C a steep increase can be observed until a peak value. Then a sharp decrease is obtained until very low values which are maintained until 200 °C. As in storage modulus graphs, GFRPs composed of unidirectional fibers in longitudinal direction have higher loss modulus than the same composites with the fibers in the transverse direction.

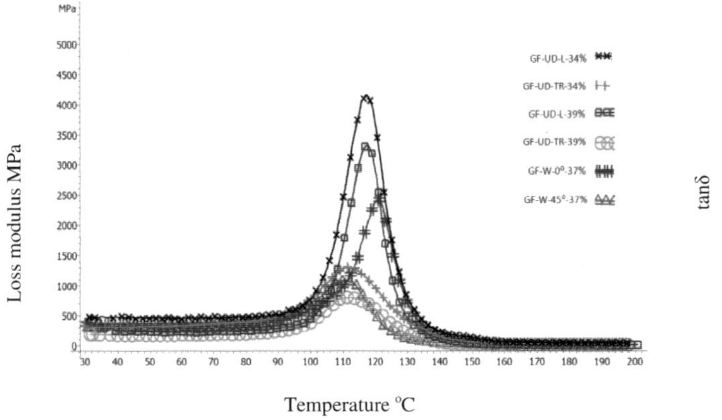

Figure 7. Loss modulus of GFRPs as function of temperature.

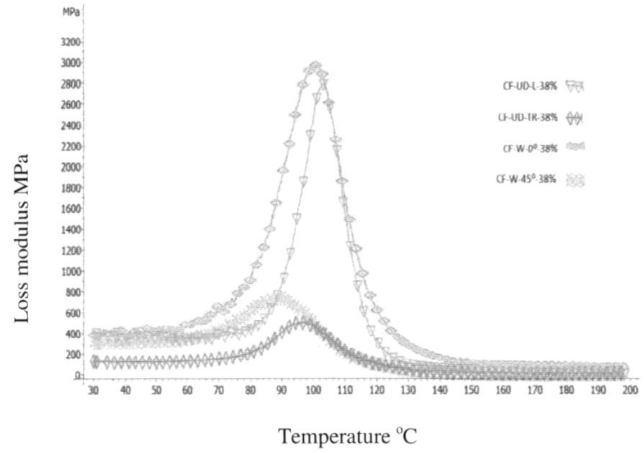

Figure 8. Loss modulus of CFRPs as a function of temperature.

3.3. tanδ

The loss factor (tanδ) is defined as the ratio of loss modulus to storage modulus:

$$\tan\delta = E''/E' \qquad (1)$$

and it is independent of geometry effects.

Figures 9 and 10 show the tanδ values of GFRPs and CFRPs as temperature increases. From these figures, it is observed that after about the 80 °C there is a rapid raise in the values of the tanδ until a peak point while then a steep decrease is observed.

In figure 10, a remarkable observation is that the curve of the CF-W-45°-38% after the peak follows a non linear decrease. This can be attributed to degradation of the specimen at high temperatures under the 3 point bending mode due to its small thickness. Moreover, a second specimen of this composite was tested and the same nonlinear curve was observed.

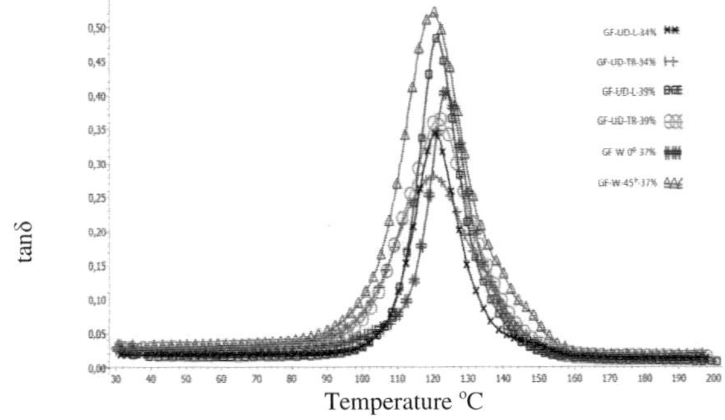

Figure 9. tanδ of GFRPs as function of temperature.

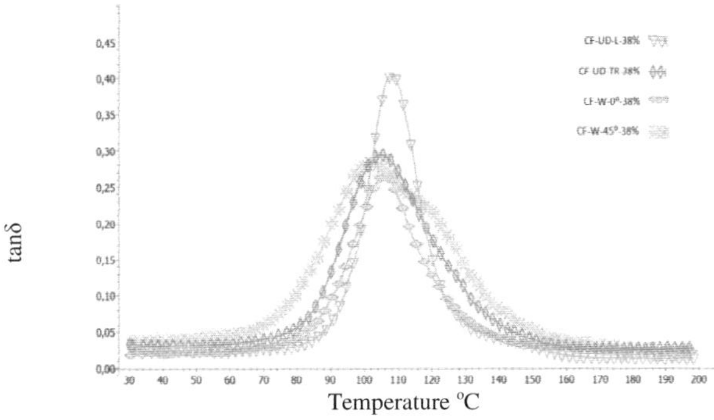

Figure 10. tanδ of CFRPs as a function of temperature.

3.4. *Glass transition temperature (Tg)*

The Tg temperature is a point for disagreement between the scientists because there are many methods for its determination. As a result, it is essential for the method which was followed for its specification to be mentioned. In this research, the Tg temperatures were obtained by two different ways; peak of loss modulus and tanδ curves respectively. These results are presented in table II. As it can be seen from a comparison between these two methods, the Tg temperatures obtained by the peak of tanδ curves are a little higher than these revealed from the loss modulus curves.

Table 2. Glass transition temperatures (Tg) °C

	Tg values obtained by loss modulus peak	Tg values obtained by tanδ peak
GF-UD-L-34%	116,91	120,54
GF-UD-TR-34%	111,69	119,85
GF-UD-L-39%	117,47	121,27
GF-UD-TR-39%	111,68	121,83
GF-W-0°-37%	120,79	123,66
GF-W-45°-37%	110,46	120,01
CF-UD-L-38%	103,08	107,54
CF-UD-TR-38%	96.27	104,96
CF-W-0°-38%	100,97	106,56
CF-W-45°-38%	88,93	101,42

4. Conclusions

During this study, GFRPs and CFRPs were produced by vacuum bag oven method and their dynamic mechanical properties were explored. Various types on fibers' architecture were used such as unidirectional fibers placed in longitudinal or transverse direction and woven fabrics oriented at 0° (fibers $0^o/90^o$) or 45° (fibers $-45^o/+45^o$).
From this viewpoint the results indicate that as the percentage of glass fibers in GFRPs increases the composites exhibit higher storage and loss modulus. Furthermore, in terms of storage and loss modulus the use of the fibers in the longitudinal direction is much more effective than if they are used in the transverse direction. In addition, very low thickness of CFRPs seems to be a restrictive factor for the successful completion of DMA temperature sweeps experiments under three point bending mode. Generally, the vacuum bag oven method is a cheap and very efficient method for the production of FRP composites by using prepreg materials.

5. Acknowledgment

This work and the project is realized with the financial support of the internal grant of TBU in Zlín No. IGA/FT/2018/004 funded from the resources of specific university research.

6. References

[1] T. Centea, L.K. Grunenfelder and S.R. Nutt, "A review of out-of-autoclave prepregs – Material properties, process phenomena, and manufacturing considerations," Composites: Part A: Applied Science and Manufacturing, Volume 70, March 2015, pp. 132-154, https://doi.org/10.1016/j.compositesa.2014.09.029

[2] R. Murugan, R. Ramesh and K. Padmanabhan. "Investigation on static and dynamic mechanical properties of epoxy based woven fabric Glass/Carbon hybrid composite laminates," Procedia Engineering, Volume 97, 2014, pp. 459-468, https://doi.org/10.1016/j.proeng.2014.12.270

[3] D. Olmos , K. Bagdi , J. Mózcó, B. Pukánszky and J. González- Benito, "Morphology and interphase formation in epoxy/PMMA/glass fiber composites: effect of the molecular weight of the PMMA," Journal of Colloid and Interface Science, Volume 360, Issue 1, August 2011, pp. 289-299, https://doi.org/10.1016/j.jcis.2011.04.028

[4] João R.Correia, Marco M. Gomes, José M. Pires and Fernando A. Branco, "Mechanical behaviour of pultruded glass fibre reinforced polymer composites at elevated temperature: experiments and model assessment," Composite Structures, Volume 98, April 2013, pp. 303-313, https://doi.org/10.1016/j.compstruct.2012.10.051

[5] Yu Bai, Nathan L. Post, John J. Lesko and Thomas Keller "Experimental investigations on temperature-dependent thermo-physical and mechanical properties of pultruded GFRP composites," Thermochimica Acta, Volume 469, Issues 1-2, March 2008, pp. 28-35, https://doi.org/10.1016/j.tca.2008.01.002

[6] R. Steigmann, A. Savin, V. Goanta. P. D. Barsanescu, B. Leitoiu, N. Iftimie, M. D. Stanciu and I. Curtu, "Determination of mechanical properties of some glass fiber reinforced plastics suitable to wind turbine blade construction," IOP Conference Series: Materials Science and Engineering, Volume 147, Issue 1, 2016, doi:10.1088/1757-899X/147/1/012140

[7] S. Cabral-Fonseca, J. R. Correia, M. P. Rodrigues and F. A. Branco, "Artificial accelerated ageing of GFRP pultruded profiles made of Polyester and Vinylester resins: characterisation of physical-chemical and mechanical damage," strain, 2012, Volume 48, Issue 2, pp. 162-173, doi: 10.1111/j.1475-1305.2011.00810.x

[8] Vasileios M. Drakonakis, James C. Seferis and Charalambos C. Doumanidis, "Curing pressure influence of out-of-Autoclave processing on structural composites for commercial aviation," Advances in Materials Science and Engineering, Volume 2013, Article ID 356824, 14 pages, doi:10.1155/2013/356824

[9] D. Kaka, J. Rongong, A. Hodzic and C. Lord, "Dynamic mechanical propertis of woven carbon fibre reinforced thermoplastic composite materials," Proc. of the 20th International Conference on Composite Materials (ICCM), July 2015

[10] Ana X. H. Yong, Graham D. Sims, Samuel J. P. Gnaniah, Stephen L. Ogin and Paul A. Smith, "Heating rate effects on thermal analysis measurement of Tg in composite materials," Advanced Manufacturing: Polymer and Composites Science, Vol 3, pp. 43-51, 2017, doi:10.1080/20550340.2017.1315908

[11] M. K. Gupta and Kunwar Rohit, "Multi layers glass fibres reinforced epoxy composite: dynamic mechanical analysis," Advanced Materials Proceedings, Volume 2, Issue 8, pp.518-520, 2017, doi: 10.5185/amp.2017/810

[12] Guillaume Kister and Eleftheria Dossi, "Cure monitoring of CFRP composites by dynamic mechanical analyser," Polymer Testing, Volume 47, October 2015, pp. 71-78, https://doi.org/10.1016/j.polymertesting.2015.08.009

[13] W.K. Goertzen and M.R. Kessler,"Dynamic mechanical analysis of carbon/epoxy composites for structural pipeline repair," Composites Part B: Engineering, Volume 38, Issue 1, January 2007, pp. 1-9 , https://doi.org/10.1016/j.compositesb.2006.06.002

[14] Andrzej Katunin, Adam Gnatowski and Wojciech Kajzer, "Evolution of static and dynamic properties of GFRP laminates during ageing in deionized and seawater," Advanced Composites Letters, Vol. 24, July 2015, pp. 38-43

IOP Conf. Series: Materials Science and Engineering **362** (2018) 012006 doi:10.1088/1757-899X/362/1/012006

Prediction of Elastic Constants of the Fuzzy Fibre Reinforced Polymer Using Computational Micromechanics

Marzena Pawlik, Yiling Lu

Department of Mechanical Engineering and the Built Environment, University of Derby, Derby, United Kingdom

e-mail: m.pawlik@derby.ac.uk, y.lu@derby.ac.uk

Abstract. Computational micromechanics is a useful tool to predict properties of carbon fibre reinforced polymers. In this paper, a representative volume element (RVE) is used to investigate a fuzzy fibre reinforced polymer. The fuzzy fibre results from the introduction of nanofillers in the fibre surface. The composite being studied contains three phases, namely: the T650 carbon fibre, the carbon nanotubes (CNTs) reinforced interphase and the epoxy resin EPIKOTE 862. CNTs are radially grown on the surface of the carbon fibre, and thus resultant interphase composed of nanotubes and matrix is transversely isotropic. Transversely isotropic properties of the interphase are numerically implemented in the ANSYS FEM software using element orientation command. Obtained numerical predictions are compared with the available analytical models. It is found that the CNTs interphase significantly increased the transverse mechanical properties of the fuzzy fibre reinforced polymer. This extent of enhancement changes monotonically with the carbon fibre volume fraction. This RVE model enables to investigate different orientation of CNTs in the fuzzy fibre model.

1. Introduction

Fibre reinforced polymers (FRP) enjoy increasing applications, and mechanical properties are key to exploit their full potential. Traditionally, mechanical properties of composites are extracted from experimental testing; however, this process is very tedious and expensive. Computational micromechanics becomes an efficient tool to predict elastic constants of FRP. This numerical approach can tailor the composition and structure of composites to meet specific requirements even before a physical sample is prepared.

Computational micromechanics analyses the composite properties at the constituent level using a representative volume element (RVE) [1]. The RVE is theoretically the smallest volume representing the whole material. Most composites have fibres randomly arranged in the microstructure; however, a good approximation of elastic constants can be achieved by periodic microstructures [2]. Generally, three types of RVE models with cylindrical [3], square [4] and hexagonal [5] fibres arrays appear in the literature. These models have been successfully applied to the determination of the FRP properties, where the proposed RVE models consist of two phases such as the fibre and matrix. Recently, the effects of an interphase have attracted a lot of interest. The interphase is defined as a thin layer between fibre and matrix which is usually created by the protective coating of the fibre and manufacturing processes [6].

Kari et al. [7] observed that overall material properties of the unidirectional lamina were significantly affected by interphase, particularly in the transverse direction. Xu et al. [8] discussed the results of volume fraction and stiffness of the interphase on the engineering elastic constants of the composite. In most works, the interphase is simplistically treated as isotropic [7-11].

Carbon nanotubes (CNTs) discovered by Iijima [12] have long been tried as the enhancer of nanocomposites because of superior and multi-functional properties and the axial Young's modulus exceeding 1 TPa [13]. In recent years, CNTs have also been applied in FRP by growing [14-17] or spraying onto the fibre surface [18]. Coated with CNTs carbon fibres are then embedded in the matrix forming the fuzzy fibre reinforced polymer (FFRP). Experimental studies found that the presence of CNTs increased not only the interfacial shear strength [19] but also material moduli of FFRP. For example, Mathur et al. [20] observed 28% increase in flexural modulus of the FFRP over the pure carbon fibre based composites.

Promising experimental results have motivated researchers to study the effects of CNTs on the FFRP properties by analytical and numerical approaches [21-28]. For example, Kulkarni et al. [21] estimated mechanical properties of a nano-reinforced laminated composite with two different volume fractions of CNTs using multiscale modelling and cylindrical RVE micromechanical method. Numerically predicted transverse modulus was 40% higher than experimentally determined value. Ray et al. [22] proposed a shear lag model to obtain the response of the CNTs coated piezoelectric fibre reinforced composites. Kundawal and Ray [23] developed a method to evaluate elastic constants of continuous fuzzy fibre composites using mechanics of materials combined with Mori-Tanaka approach. Whereas, Chatzigeorgiou et al. [24] characterised fuzzy fibre reinforced composites considering various CNTs volume fractions and lengths. Ren et al. [25] experimentally and computationally estimated the piezoresistivity of fuzzy fibre reinforced polymer by 3-D FEM multiscale model. More recently, Rafiee and Ghorbanhosseini [26] applied a stochastic multi-scale model to obtain mechanical properties of FFRP with randomly orientated CNTs on the fibre surface.

In the present work, the RVE with hexagonal fibres array is used for numerical simulation to predict the properties of the fibre reinforced polymer with the concept of the CNTs reinforced interphase. The influence of the interphase properties on the response of the FFRP is analysed for a various carbon fibre volume fractions. Two types of models: with and without the CNTs reinforced interphase are compared to highlight its influence on material properties.

2. RVE method

2.1. Fuzzy fibre polymer – RVE model.

Fig.1a shows a schematic representation of a unidirectional carbon fibre reinforced lamina. The mechanical behaviour of this composite can be effectively studied using the hexagonal RVE model as presented in Fig.1 b. The RVE model contains three phases: the carbon fibre, the CNTs reinforced interphase and the epoxy resin. In the fuzzy fibre, CNTs are radially grown on the surface of the carbon fibre (Fig.2a); therefore, the CNTs reinforced interphase is composed of the nanotubes and epoxy. The transversely isotropic features of CNTs are taken into account in the reinforced interphase. Their material properties symbols and subscripts follow the local coordinate system 1'-2'-3' presented in Fig 2b. Fuzzy fibres are arranged in the unidirectional lamina and follow Cartesian coordinate system with axes denoted by 1-2-3, where, 1 is the carbon fibre direction.

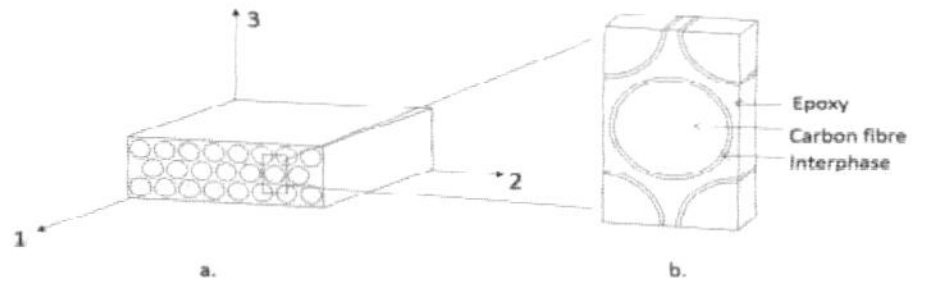

Figure 1. Schematic representations of a. the carbon fibre reinforced polymer, b. the three-phase RVE.

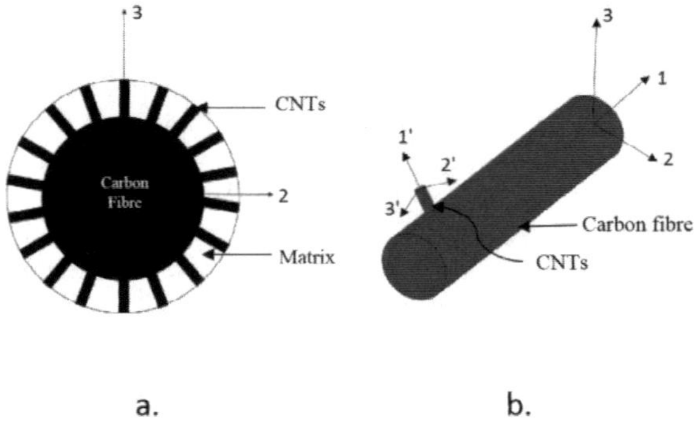

Figure 2. Schematic illustrations of a. cross-section view of the fuzzy fibre, b.perspective view of the fuzzy fibre.

2.2. Material properties of each phase

Two-phase and three- phase models are investigated to highlight the effects of the CNTs-reinforced interphase on the mechanical properties of the FFRP. In the two-phase model, mechanical properties of the transversely isotropic T650 carbon fibre and the isotropic EPIKOTE 862 epoxy resin are selected (Table I) [24]. Properties of the transversely isotropic carbon fibre are assigned according to the lamina coordinate system. Dimensions of the RVE are calculated based on the fibre volume fraction and fibre diameter (5 μm). For example, for 25 % of carbon fibre volume fraction, height, width and thickness are $8.23, 4.76$ and 1.19 μm respectively. Detailed information can be found in [1]. The three-phase model also includes the CNTs reinforced interphase. The "VEORIENT" command is applied to the RVE model to define the orientation of the CNTs. This command allows specifying element orientation to control transversely isotropic material property direction. The CNTs reinforced interphase has a thickness of 2 μm. The average volume fraction of CNTs in the interphase amounts to 42% [24]. As a result of CNTs mechanical characteristics and orientation, the interphase is transversely isotropic. The material properties of the CNTs interphase are taken from literature. Poisson's ratios are calculated from the bulk moduli provided by [24] (Table II).

Table 1. Material properties of the carbon fibre and epoxy resin.

Matrix : EPIKOTE 862 Resin	
Modulus, $\boldsymbol{E_m}$	3 GPa
Poisson's ratio, v_m	0.3
Fibre: Carbon fibre T650	
Longitudinal modulus, E_{f1}	241 GPa
Transverse modulus, E_{f2}	14.5 GPa
In-plane shear modulus, G_{f12}	22.8 GPa
Transverse shear modulus, G_{f23}	4.8 GPa
In-plane Poisson's ratio, v_{f12}	0.27

Matrix : EPIKOTE 862 Resin	
Modulus, $\boldsymbol{E_m}$	3 GPa
Diameter	5 µm

Table 2. Material properties of the CNTs reinforced interphase.

Interphase : CNTs reinforced interphase	
Longitudinal modulus, E_1,	298.64 GPa
Transverse modulus, E_2,	7.01 GPa
In-plane shear modulus, G_{12},	2.81 GPa
Transverse shear modulus, G_{23},	2.52 GPa
In-plane Poisson's ratio, v_{12},[a]	0.1
Out-of-plane Poisson's ratio , v_{23},	0.29

a. estimated property

Both models are simulated using ANSYS APDL 17.2. Perfect bonding conditions between each phase are assumed. The meshes generated for the models investigated are 20-node 3-D solid elements (SOLID186). Mesh independent test was accomplished to eliminate the influence of the number of elements on the accuracy of the results. The whole meshed model contains approximately 7000 elements.
Keeping the same properties and dimensions for all constituents, we performed numerical simulations where carbon fibre volume fractions ranging from 0 to 25% for two and three-phase models. Note that 25% carbon fibre volume fraction corresponds to 81% volume fraction of the FFRP model (carbon fibre together with the CNTs interphase).

2.3. Evaluation of elastic constants

The prediction of elastic constants of the fibre-reinforced composite is based on Hooke's law for transversely isotropic materials (1), where C_{ijkl} is a stiffness matrix, and $\bar{\sigma}_{ij}$ and $\bar{\varepsilon}_{ij}$ are the volume average stresses and strains respectively. The volume average strains and stresses are obtained by (2), where V is the volume of the RVE.

$$\begin{Bmatrix} \bar{\sigma}_1 \\ \bar{\sigma}_2 \\ \bar{\sigma}_3 \\ \bar{\tau}_4 \\ \bar{\tau}_5 \\ \bar{\tau}_6 \end{Bmatrix} = \begin{bmatrix} C_{11} & C_{12} & C_{12} & 0 & 0 & 0 \\ C_{12} & C_{22} & C_{23} & 0 & 0 & 0 \\ C_{12} & C_{23} & C_{22} & 0 & 0 & 0 \\ 0 & 0 & 0 & \frac{1}{2}(C_{22} - C_{23}) & 0 & 0 \\ 0 & 0 & 0 & 0 & C_{66} & 0 \\ 0 & 0 & 0 & 0 & 0 & C_{66} \end{bmatrix} \begin{Bmatrix} \bar{\varepsilon}_1 \\ \bar{\varepsilon}_2 \\ \bar{\varepsilon}_3 \\ \bar{\gamma}_4 \\ \bar{\gamma}_5 \\ \bar{\gamma}_6 \end{Bmatrix} \tag{1}$$

The periodic boundary conditions are applied to the RVE model to compute stresses, strains and calculate the overall elastic matrix, C_{ijkl}. The comprehensive description of the periodic boundary conditions can be found elsewhere [1, 2, and 29].

$$\bar{\varepsilon}_{ij} = \frac{1}{V}\int_V \varepsilon_{ij} dV$$

$$\bar{\sigma}_{ij} = \frac{1}{V}\int_V \sigma_{ij} dV \tag{2}$$

When the components of the tensor C are known, the five elastic properties of the homogenised material can be calculated (3); where E_1 and E_2 represent the longitudinal and transverse moduli, v_{12} and v_{23} are the in-plane and out-of-plane Poisson's ratios, G_{12} is the in-plane shear modulus, and G_{23} is out-of-plane shear modulus. Calculated mechanical properties for both models are compared and presented in Fig. 3 to 6.

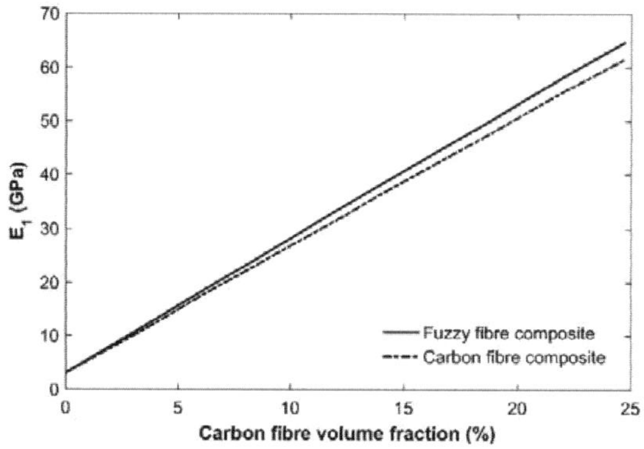

Figure 3. Longitudinal modulus versus carbon fibre volume fraction.

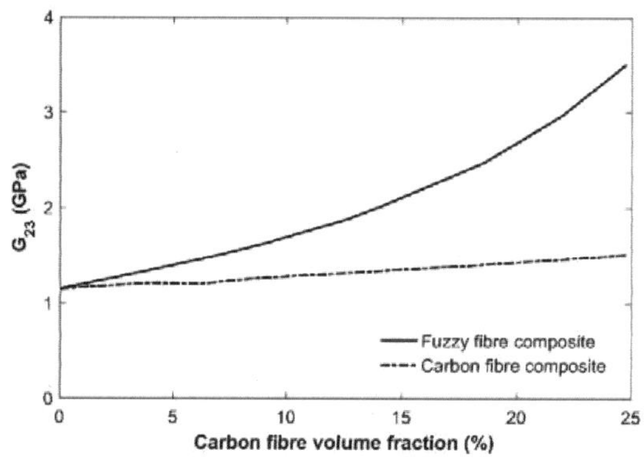

Figure 4. Transverse modulus versus carbon fibre volume fraction.

$$
\begin{gathered}
E_1 = C_{11} - 2C_{12}^2/(C_{22} + C_{23}) \\
E_2 = [C_{11}(C_{22} + C_{23}) - 2C_{12}^2](C_{22} - C_{23})/(C_{11}C_{22} - C_{12}^2) \\
v_{12} = C_{12}/(C_{22} + C_{23}) \\
v_{23} = [C_{11}C_{23} - C_{12}^2]/(C_{11}C_{22} - C_{12}^2) \\
G_{12} = C_{66}
\end{gathered} \tag{3}
$$

3. Results

3.1. Longitudinal modulus

Fig.3 presents longitudinal modulus, E_1,versus carbon fibre volume fraction of fuzzy fibre and carbon fibre composites. In both cases, for 0% carbon volume fraction results started from 3 GPa, which was the modulus of the pure matrix. For 25% of carbon volume fraction in the fuzzy fibre material, the

IOP Conf. Series: Materials Science and Engineering **362** (2018) 012006 doi:10.1088/1757-899X/362/1/012006

longitudinal modulus increased up to 65 GPa. For the same volume fraction of composite without the interphase, the modulus was 61.7 GPa. The nearly linear relationship between modulus and volume fraction was obtained for both models. Results for two and three-phase models were close to each other.

3.2. Transverse modulus

Transverse modulus, E_2 ,for the two-phase model was almost insensitive to the increment of carbon fibre volume fraction (Fig.4). However, the significant increase was observed for the FFRP. Transverse modulus increased from 3 GPa, for 0% volume fraction, up to 10.4 GPa for 25% of carbon fibre volume fraction. In comparison to the two-phase composite, the presence of CNTs interphase increased transverse modulus more than 200%, when volume fraction reached above 20%. The improved performance of FFRP is attributed to the high axial stiffness of the CNTs which are orientated in the transverse direction of the fibre.

3.3. Shear moduli

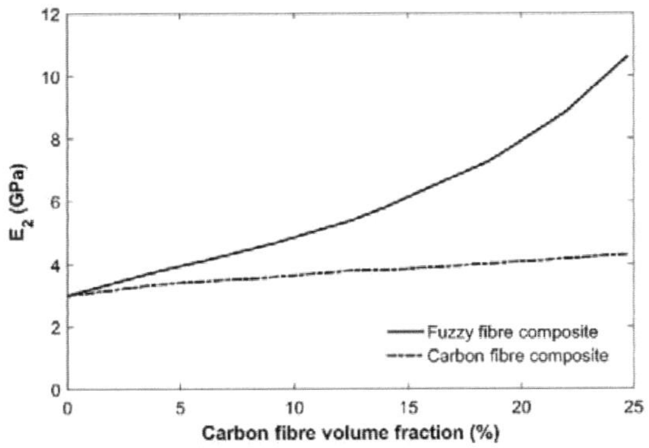

Figure 5. Out-of-plane shear modulus versus carbon fibre volume fraction.

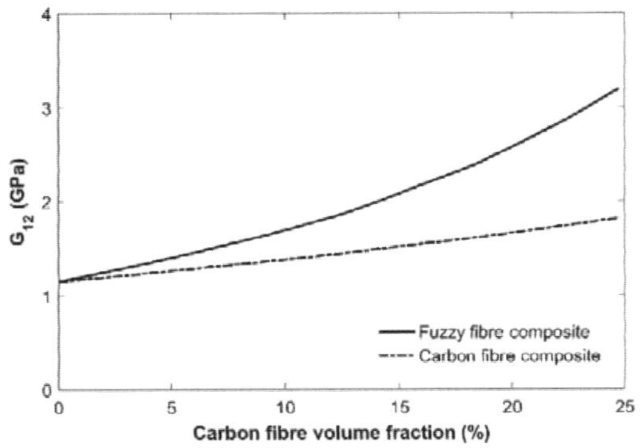

Figure 6. In-plane shear modulus versus carbon fibre volume fraction.

Both similar curves for in-plane and out-of-plane shear moduli were observed in Fig.5 and 6. Shear moduli, G_{12} and G_{23}, values for carbon fibre model were nearly insensitive to the change of the carbon fibre volume fraction. Significant increases were observed in the case of the fuzzy fibre. Up to 10 % of carbon fibre volume fraction, the increase was almost linear. Above 10%, shear moduli grow exponentially. For 25 % of carbon fibre volume fraction, shear moduli enhanced more than twice.

4. Discussion

Numerical results predicted by the hexagonal RVE perfectly match analytical results calculated by composite cylinders method [24], which are not shown on the graphs. Kundawal et al. [23, 27] established similar trends in carbon fibre volume fraction, and effective elastic coefficients of fuzzy fibre reinforced composites predicted by mechanics of materials combined with Mori-Tanaka approach.

The purpose of coating fibres with nanofillers is to strengthen the interphase between the fibre and matrix. Numerical results show a significant improvement of properties in the transverse direction. Transverse properties of the carbon fibre-reinforced composites are governed by matrix-dominant properties and are considerably lower in comparison to longitudinal. The presence of CNTs in the matrix surrounding the carbon fibre enhances the performance of the FFRP. This enhancement is attributed to the unique orientation where axial direction of the CNT is aligned with the transverse direction of the carbon fibre. An experimental study [13] supports these findings. According to Kulkarni et al. [21], the transverse elastic modulus can usually be improved about three times with respect to the value of the pure matrix. In this study, for 25 % carbon fibre volume fraction, the transverse modulus increased more than twice. The achieved increase depends on the volume fraction of the CNTs in the interphase and also on the quality of manufacturing process.

Contrarily, the presence of the CNTs radially grown on the surface of carbon fibre has negligible influence on the longitudinal modulus of the fuzzy fibre composite. The highest improvement of longitudinal modulus is observed for the 25% carbon fibre volume fraction and results in only 5% increase. This lower increase is associated with weaker properties of the interphase in the carbon fibre direction (1-direction in global coordinate system) which is caused by lower transverse properties of the single nanotube.

In the case of fuzzy fibre, cylindrical RVE is more commonly used due to the axisymmetric nature [21]. However, the hexagonal RVE model frees us from the cylindrical coordinate system which is required by the analytical way. This model enables to study a various distribution and directions of nanofillers, for example, the graphene sheets parallel aligned to the fibre surface or randomly distributed in the reinforced interphase [28].

The main limitation of this study is the assumption of perfectly straight CNTs, evenly distributed on the fibre surface. In reality, during the manufacturing process, some defects may occur. Moreover, the thickness of CNTs interphase may not be uniform throughout the fibre. These problems will be considered in the future work.

5. Conclusion

To sum up, our work has investigated the effects of the CNTs interphase on the fuzzy fibre composite using computational micromechanics. The RVE with hexagonal fibre array successfully predicted elastic constants of the fuzzy fibre composite. The results indicate that CNTs have a significant influence on transverse properties of the FFRP. This study provides the framework to study the fibre-reinforced composite with the existence of the nano-reinforced interphase. This numerical model allows dealing with arbitrary shapes of fibres which cannot always be tackled analytically. The developed code can simply and promptly calculate the required properties. Direction-dependent properties of the interphase and fibres can be readily implemented.

6. Acknowledgment

Marzena Pawlik is supported by graduate teaching assistant funding from College of Engineering and Technology at Derby University.

7. References

[1] EJ.Barbero,Finite Element Analysis of Composite Materials Using ANSYS, Second Edition, CRC Press, 2014.

[2] D.Trias,J.Costa,J.A.Mayugo,J.E.Hurtado,"Random model versus periodic periodic models for fibre reinforced composites", Computational Materials Science, vol. 38(2),2006, pp.316-324, doi:10.1016/j.commatsci.2006.03.005.
[3] Z.Hu,R.Karki,"Prediction of mechanical properties of three-dimensional fabric composites reinforced by transversely isotropic carbon fibers,",Journal of Composite Materials, vol.49(12),2015, pp.1513-1524, doi.org/10.1177/0021998314535960.
[4] C.T.Sun and R.S.Vaidya,"Prediction of composite properties from a representative volume element,"Composite Science and Technology, vol.56(2), 1996,pp.171-179, doi:10.1016/0266-3538(95)00141-7.
[5] J.H.Lim,M.Henry,D.S.Hwang,D.Sohn,"Numerical prediction of fiber mechanical properties considering random microstructures using inverse analysis with quasi-analytical gradients,"Applied Mathematics and Computations,vol.273,2016,pp.201-216, doi:10.1016/j.amc.2015.09.076.
[6] J.K.Kim,Y.W.Mai,Engineered interfaces in fiber reinforced composites, Elsevier,1998.
[7] S.Kari,H.Berger,U.Gabber,R.Guinovart-Diaz,J.Bravo-Castillero,R.Rodriguez,Ramos,"Evaluation of influence of interphase material parameters on effective material properties of three phase composites,"Composite Science and Technology,vol.68,2008,pp.648-691, doi:10.1016/j.compscitech.2007.09.009.
[8] Y.Xu,S.Du,J.Xiao,Q.Zhao,"Evaluation of the effective elastic properties of long fiber reinforced composites with interphases,"Computational Material Science,vol.61,2012,pp.43-41, doi:10.1016/j.commatsci.2012.03.048.
[9] P.Kumar,R.Chandra,S.P.Singh,"Effective elastic constants of fiber-reinforced polymer-matrix composites with the concept of the interphase,"Composites Interfaces, vol.17,2010,pp.15-35, doi:10.1163/092764409X12580201111502.
[10] G.Wacker,K.Bledzki,A.Chateb,"Effect of interphase on the transverse Young's modulus of glass/epoxy composites," Composite Part A,vol.29,1998,pp.619-626, doi:10.1016/S1359-835X(97)00116-4.
[11] Y.Liu,Y.Yan,H.She,M.He,"Prediction on macroscopic elastic properites of interphase-contained long-fiber-reinforced composites and multiple nonlinear regression analysis,"Journal of Reinforeced Plastics and Composites, vol.31, 2012, pp.1143-1157, doi:10.1177/0731684412454461.
[12] S.Ijima,"Helical microtubules of graphitic carbon," Nature, vol.354, 1991, pp.56-58, doi:10.1038/354056a0.
[13] Yu, M. F., Files, B. S., Arepalli, S., & Ruoff, R. S. "Tensile loading of ropes of single wall carbon nanotubes and their mechanical properties,"Physical review letters, vol.84(24), 2000,pp.5552-5, https://doi.org/10.1103/PhysRevLett.84.5552.
[14] R.J.Sager,P.J.Klein,D.C.Lagoudas,Q.Zhang,J.Liu,L.Dai,et.al,"Effect of carbon nanotubes on the interfacial shear strength of T650 carbon fiber in an epoxy matrix,"Composite Science and Technology,vol.69,2009,p.898-904, doi:10.1016/j.compscitech.2008.12.021.
[15] Wood, C. D., Palmeri, M. J., Putz, K. W., Ho, G., Barto, R., & Brinson, L. C.,"Nanoscale structure and local mechanical properties of fiber reinforced composites containing MWCNT-grafted hybrid glass fibers," Composites Science and Technology, vol.72(14),2012,pp. 1705-1710, doi.org/10.1016/j.compscitech.2012.06.008.
[16] Zhao, Z. G., Ci, L. J., Cheng, H. M., & Bai, J. B.," The growth of multi-walled carbon nanotubes with different morphologies on carbon fibers,"Carbon,vol.43(3),2005,pp.663-665, doi.org/10.1016/j.carbon.2010.06.006.
[17] S.Y.Jin,R.J.Young,S.J.Eichhorn,"Hybrid carbon fibre-carbon nanotube composite interfaces," Composite Science and Technology, vol.95,2014,pp.114-120, doi:10.1016/j.compscitech.2014.02.015.

[18] F.L.Shan,Y.Z.Gu,M.Li,Y.N.Liu,Z.G.Zhang,"Effect of deposited carbon nanotubes on interlaminar composites using a developed spraying processing,"Polymer Composites,vol.34,2013,pp.41-50, doi:10.1002/pc.22375.

[19] Thostenson, E. T., Li, W. Z., Wang, D. Z., Ren, Z. F.,Chou, T. W., "Carbon nanotube/carbon fiber hybrid multiscale composites, "Journal of Applied physics, vol.91(9),2002,pp.6034-6037, doi.org/10.1063/1.1466880.

[20] Mathur, R. B., Chatterjee, S., & Singh, B. P.,"Growth of carbon nanotubes on carbon fibre substrates to produce hybrid/phenolic composites with improved mechanical properties," Composites Science and Technology, vol.68(7), 2008,pp.1608-1615, doi.org/10.1016/j.compscitech.2008.02.020.

[21] M.Kulkarni,D.Carnahan,K.Kulkarni,D.Qian,J.L.Abot,"Elastic response of carbon nanotube fiber reinforced polymeric composite: A numerical and experimental study," Composite Part B,vol.41,2010,pp.414-421, doi:10.1016/j.compositesb.2009.09.003.

[22] M.C.Ray,"A shear lag model of Piezoelectric composite reinforced with carbon nanotubes-coated Piezoelectric fibers," Internationa Journal of Mechanics and Materials in Design,vol.6(2),2010,pp.147-155, doi:10.1007/s10999-010-9118-2.

[23] S.I.Kundalwal,M.C.Ray,"Micromechanical analysis of fuzzy fiber reinforced composites,"International Journal of Mechanics and Materials in Design,vol.7,2011,pp.149-166, doi:10.1007/s10999-011-9156-4.

[24] G.Chatzigeorgiou,G.D.Seidel,D.C.Lagoudas,"Effective mechanical properties of "fuzzy fiber" composites,"Composite Part B,vol.43,2012, pp.2577-2593, doi:10.1016/j.compositesb.2012.03.001.

[25] X.Ren,J.Burton,G.D.Seidel,K.Lafdi,"Computational multiscale modeling and characterization of piezoresistivity in fuzzy fiber reinforced polymer composites,"International Journal of Solids and Structures,vol.54,pp.121-134, doi:10.1016/j.ijsolstr.2014.10.034.

[26] R.Rafiee,A.Ghorbanhosseini,"Stochastic multi-scale modeling of randolmy grown CNTs on carbon fiber,"Mechanics of Materials,vol.106,pp.1-7, doi:10.1016/j.mechmat.2017.01.001.

[27] S.I.Kundalwal,M.C.Ray,"Effective properties of a novel continuous fuzzy-fiber reinforced composite using the method of cells and the finite element method,"European Journal of Mechanics-A/Solids,vol.36,2012,pp.191-203, doi:10.1016/j.euromechsol.2012.03.006.

[28] M.Pawlik,A.Dean,Y.Lu,"The numerical investigation of graphene-reinforced interphase effects on the mechanical properties of carbon fibre/epoxy composites," Proc. International Conference on Composite Structure, (ICCS20),Societa Editrice Esculapiom, Sep.2017,pp.194-195,doi: 10.15651/978-88-938-5041-4.

[29] R.Luciano,E.Sacco," Variational methods for the homogenization of periodic heterogeneous media," European Journal of Mechanics-A/Solids,vol.17(4),1998,pp.599-617,doi.org/10.1016/S0997-7538(99)80024-2.

IOP Conf. Series: Materials Science and Engineering **362** (2018) 012007 doi:10.1088/1757-899X/362/1/012007

Laser Annealing on the Surface Treatment of Thin Super Elastic NiTi Wire

S Samal[1], L Heller[1], J Brajer[2], O Tyc[1], L Kadrevek[1] and P. Sittner[1]

1 Institute of Physics of the Czech Academy of Sciences, 182 21 Praha, CZ

2 Department of Machining, Czech Technical University in Prague, 166 07 Praha, CZ
E-mail: samal@fzu.cz

Abstract. Here the aim of this research is annealing the surface of NiTi wire for shape memory alloy, super-elastic wire by solid state laser beam. The laser surface treatment was carried out on the NiTi wire locally with fast, selective, surface heat treatment that enables precisely tune the localized material properties without any precipitation. Both as drawn (hard) and straight annealing NiTi wire were considered for laser annealing with input power 3 W, with precisely focusing the laser beam height 14.3 % of the Z-axis with a spot size of 1 mm. However, straight annealing wire is more interest due to its low temperature shape setting behavior and used by companies for stent materials. The variable parameter such as speed of the laser scanning and tensile stress on the NiTi wire were optimized to observe the effect of laser response on the sample. Superelastic, straight annealed NiTi wires (d: 0.10 mm) were held prestrained at the end of the superelastic plateau (ε: 5 ~6.5 %) above the superelastic region by a tensile machine (Mitter: miniature testing rig) at room temperature (RT). Simultaneously, the hardness of the wires along the cross-section was performed by nano-indentation (NI) method. The hardness of the NiTi wire corresponds to phase changes were correlated with NI test. The laser induced NiTi wire shows better fatigue performance with improved 6500 cycles.

1. Introduction

NiTi based shape memory alloys are used in a wide variety of engineering applications [1] due to their unique thermomechanical functional properties. Shape memory alloys (SMAs) belong to the category of smart materials offers the unique properties of super-elasticity and shape memory effect [2]. The near equi-atomic NiTi alloy is emerging as one of the most demanding among SMAs alloys due to the stable performance in fatigue performance. Fatigue limits the long term safe use of the material and reduces the durability of the application. Generally, fatigue occurs in the cycle of loading and unloading of the sample during application, at various operating environments (specially temperature as one of the crucial parameters during application stage) with proper surface conditions. According to theoretical prediction the solid state conversion in NiTi alloy martensite (monoclinic, B_{19}') daughter phase stable at low temperature derived from austenite (B_2) parent phase (cubic) stable at high temperature towards could deliver millions of mechanical or thermomechanical cycles as the function of strain or stress [3]. However, in practice to achieve the structural fatigue performance for NiTi alloy is a challenge to researchers. The main reason for the obstacles is the surface oxide layer formation during conventional heating method [4] of NiTi alloy for the phase transformation during heat treatment in a furnace at the range of 400 to 500°C for 30 minutes. To improve the heating process, the conventional method shifts towards laser technology that not only improves the temperature profile also avoids the oxide layer formation on the surface of NiTi wire. To improve the surface

treatment process, earlier researchers were used a various alternative process such as electric pulse heating under mechanical constraint [5] method in replace of the conventional heating method. Although electric pulse treatment is a faster process and improves the functional properties of NiTi alloy, however the upscaling of this technology is an obstacle.
Laser technique offers a lot of promising factors in surface treatment such as high throughput, rapid technology, flexibility in process technology, quick and reproducibility process with better quality in the product. Laser heating is emerging a new innovative approach for the surface treatment of NiTi wire rather than the application limited in areas of melting and cutting technology [6]. Thin films of amorphous NiTi alloys were annealed with a laser beam to improve the fatigue life with small spots of SMA [7]. However, there is lack of work presented in the areas of laser surface annealing of NiTi wire at constant strain by inducing homogenous functional properties.
In the present work, the laser annealing was carried out on the surface of NiTi wire on martensite phase (straight annealed) at constant strain. Superelastic, straight annealed NiTi wires (d: 0.10 mm) were held prestrained at the end of the superelastic plateau (ε: 5 ~6.5 %) above the superelastic region by a tensile machine (Mitter: miniature testing rig) at room temperature (RT). The laser plasma creates much larger pressure that resulted from the momentum impulse generated by surface heating rapidly by the laser pulse. As a result, laser pulse interaction generates stress waves intensity from the interaction point to the core of the alloy. The laser annealing of super-elastic, straight annealed wire was performed with the various speed of laser scanning. As a result, it can improve the annealing behavior of the sample during laser treatment and improve the fatigue cycle. The laser annealed NiTi wire (as drawn, hard) was investigated by the tensile machine. Thin NiTi wire was investigated in cyclic tensile tests beyond the end of the transformation plateau at RT until a failure happens. The aim was to generate compressive stress on the NiTi wire by the laser low temperature shape setting process to improve fatigue life. The hardness, Young's modulus and remnant depth of the laser treated NiTi wire were measured by the NI test.

2. Experimental method

2.1. Materials
NiTi superelastic wires, Ti-50.9 at.% Ni (both as drawn and straight annealing, 100 μm of diameter) were supplied by Fort Wayne Metals company.

2.2. Laser Technique
The laser experiments were carried out on the Lumonics Laser JK701H, 1064 nm solid state laser (Nd: YAG) from the Research Center for Engineering Production Technology and Technology at Czech Technical University. The continuous laser scanning was performed at a frequency of 500 Hz with a residence time (Δt) of 0.5 ms, input power 3 ~4 W.

2.3. Methods
Laser experiments were performed on the surface of NiTi wire at the tensile machine (miniature tensile machine, Metter, Max. Load: 10 N) strained with a fixed value (5~ 6 %) after plateau region. Tensile stress response of NiTi wire without and with laser treatment was tested as the function of strain. The hardness measurement was performed on laser treated NiTi wire mounted on the epoxy resin with capillary fixed at the end point by nano indentation method. Fatigue tests were carried out in strain controlled mode (strain rate: 0.01 s^{-1}) for laser annealed samples at constant temperature 20 °C. The limit values were chosen with stress value of σ_{min}: 10 MPa and σ_{max}: 700 MPa and strain values are ε_{min}: 0 % for both NiTi wire and laser annealed NiTi wire. The upper limit εmax for NiTi wire was chosen 6.5 % and 6 % for laser treated NiTi wire considered in the tensile mode of fatigue cycle until a failure happens.

3. Results

3.1. Surface treatment of as drawn sample from hard to super elastic behavior with laser modification
The hard samples of NiTi wire were fixed at Metter tensile machine with 220 MPa applied stress at a fixed position. Laser scanning was carried out on the upper surface of hard wire at a constant speed. The super-elastic behavior was observed after laser treatment on the hard wire. This experiment confirms the functional behavior of hard wire after laser shape setting. Fig. 1. displays the stress versus strain curve for hard NiTi wire before and after laser treatment. The plateau behavior explains the existence of super-elastic behavior in NiTi wire. Based on this laser surface treatment parameter were chosen for super-elastic straight annealed NiTi wire samples.

3.2. Laser annealing of super-elastic, straight annealed wire at martensitic phase by laser heating
The straight annealed, super-elastic sample was strain up to 6 % and fixed above the plateau region in the tensile machine before laser treatment. Laser heating was performed across the wire in martensitic phase in tension stage and the effect was studied carefully to induce the temperature gradient zone across the cross section from laser annealed point from the edge towards the core of the NiTi wire. With the target maintain the core of the wire unaltered as same as raw NiTi wire. Only induce the localized laser heating effect in the extended wire, as a result, to create compressive stresses from the surface region. Fig. 2. shows the similar mechanical response of NiTi wire at speed of 450 mm/min at 800 MPa and stress versus strain curves coincide with each other.

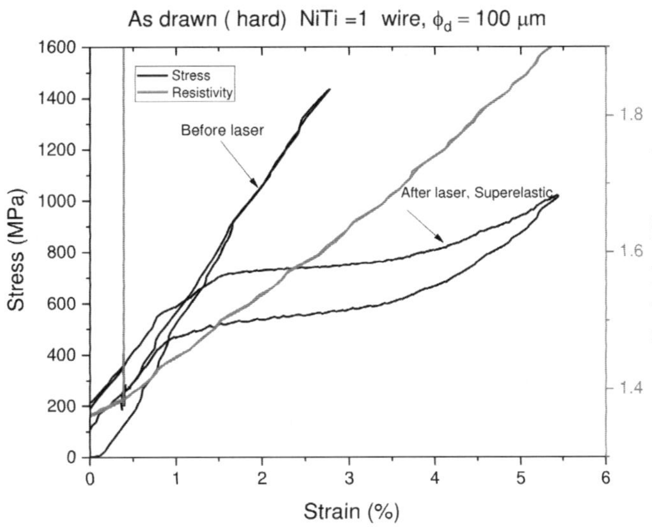

Fig. 1. Stress versus strain for hard NiTi wire before and after laser surface treatment shows response before (straight*, hard) and after laser heating (plateau, super elastic*).

However, on decreasing the speed of 370 mm/min, the stress curve reduces and hysteresis curve lowers with decreasing trend from the original response. This behavior of laser annealing sample is depicted in Fig. 3.

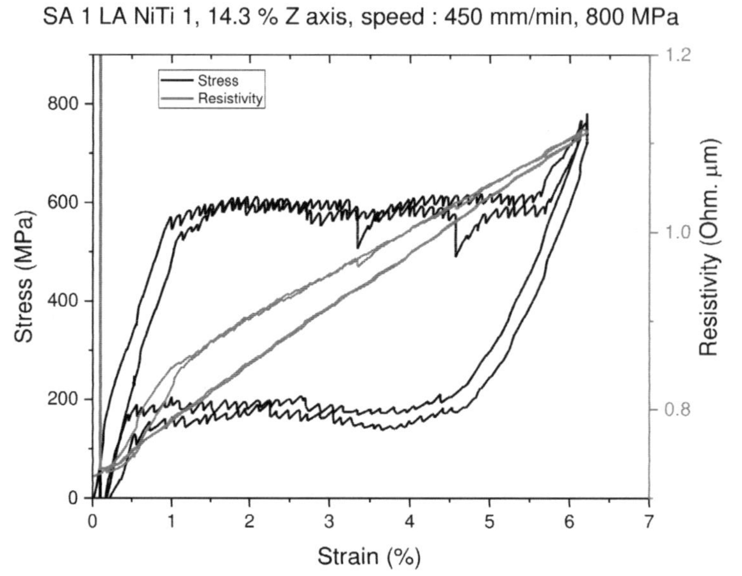

Fig. 2. Laser annealed straight annealed wire at 800 MPa with speed 450 mm/min*.

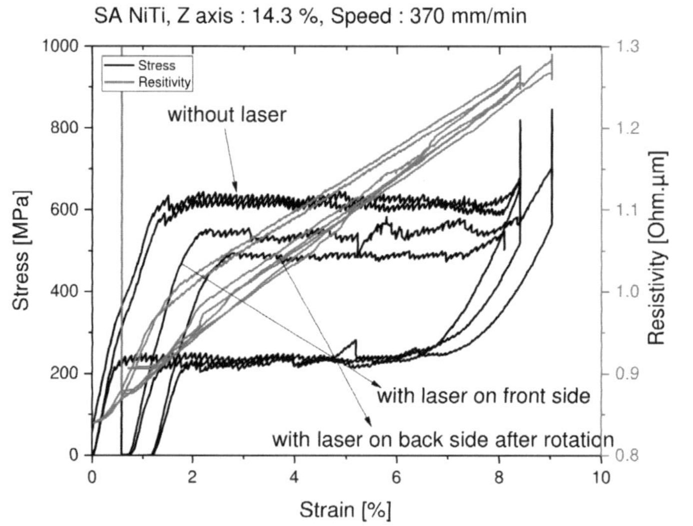

Fig. 3. Laser annealed straight annealed wire at 800 MPa with speed 370 mm/min*.

Fig. 4 represents the stress strain behavior of the sample at a lower speed of 230 mm/min. This shows laser annealed sample accumulates transformation strain more of 2 %. To interpret this mechanism of laser annealing is investigated by NI method of laser annealed NiTi wire along the cross section of the sample during loading and unloading at maximum load.

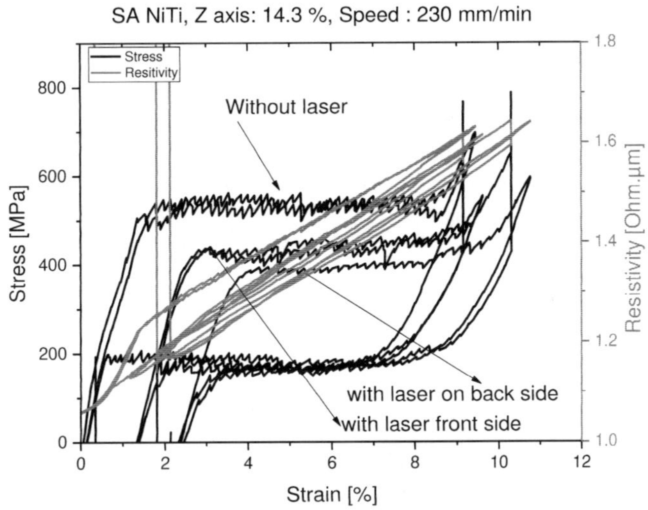

Fig. 4. Laser annealed straight annealed wire at 800 MPa with speed 230 mm/min*.
(* Noise in the curve is due to the error of manual tensile machine)

3.3. NI method of the NiTi wire without and with laser annealed samples

Table 1 shows the summary of NI results for hard NiTi wire, SA wire, and the respective laser annealed samples. Hard NiTi wire shows the higher hardness of 5.5 GPa, however, SA wire is softer shows 3.4 GPa. Laser annealed hard wire shows super-elastic behavior from the Fig. 1, reveal the hardness of 4.6 GPa, however laser annealed SA wire shows slightly increase in hardness 4.5 GPa than the raw SA (without laser) wire. On deceasing the scan speed to 370 mm/min, the annealed of SA wire shows the better result of hardness value of 3.9 GPa. On going to more lower speed of 230 mm/min, the sample shows the slightly lower value of hardness decrease the value of 3.8 GPa.

Table 1. Nano indentation test on the summary of the samples both hard and straight annealing and their respective laser treated samples.

NiTi wire / Property	Hard	SA	HARD + LA, 450 mm/min	SA + LA, 450 mm/min	SA + LA, 370 mm/min	SA + LA, 230 mm/min,
Hardness (H,GPa)	5.46 ± 0.30	3.40 ± 0.11	4.6± 0.3	4.52 ± 0.22	3.89 ± 0.19	3.84 ± 0.16
E (GPa)	78.41± 3.30	49.51 ± 2.11	59.61±3.19	50.13 ± 4.69	56.89 ± 4.98	53.76 ±1.6
Remnant depth, H eff(nm)	211.57± 5.64	273.69 ± 3.64	NA	244.39 ± 8.19	250.65 ± 5.48	254.21 3.63

3.4. Fatigue life of Straight annealed NiTi wire before and after laser treatment

Fatigue life of SA, NiTi wire shows 3150 cycles until a failure happens (Fig. 5). However, laser treated the sample with a similar response as SA, raw wire from the stress-strain response from Fig. 1 shows the slightly lower value of 3073 cycles of fatigue life until rupture. The laser annealed NiTi wire at speed of 370 mm/ min with reduced stress value shows improved fatigue life of 6500 cycles (Fig. 6). However, the functional behavior of the cold work wire after electrical pulse heating shows compare result on the previous work by the researchers [8-10].

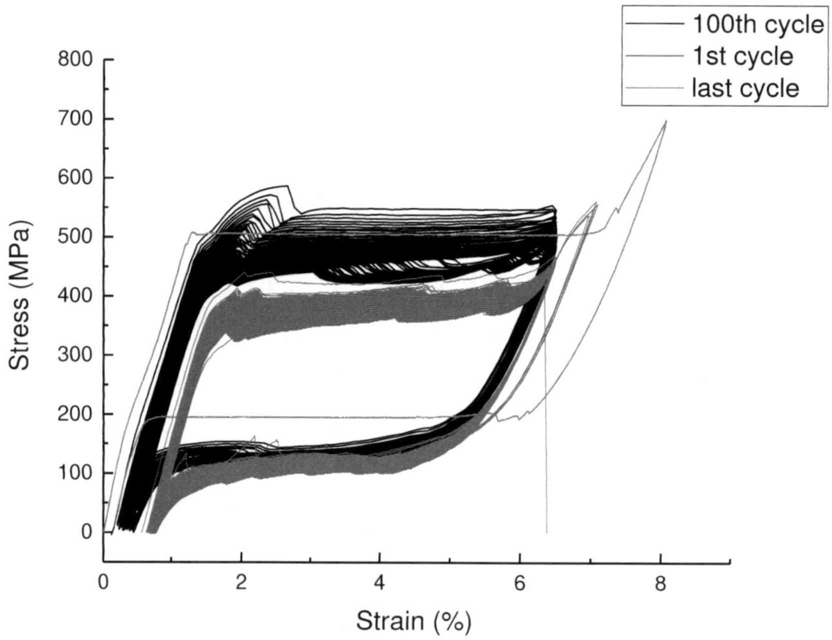

Fig. 5. SA wire (raw, without laser) shows fatigue of 3150 cycles until failure.

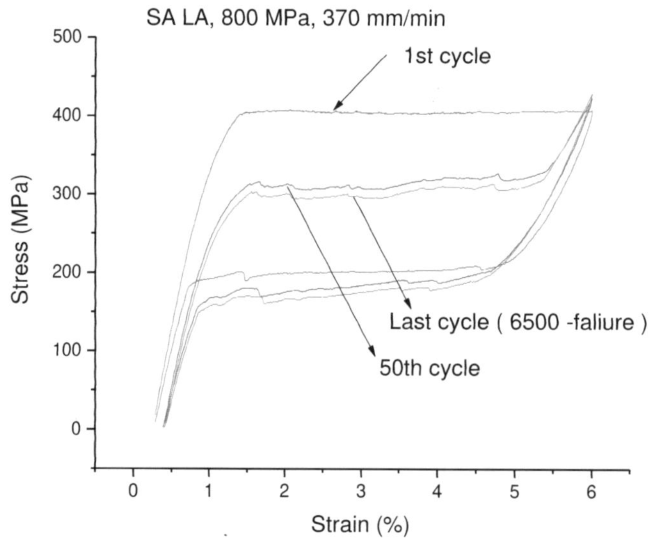

Fig. 6. The laser annealed SA wire corresponds to Fig. 4 shows the fatigue of 6500 cycles until failure.

4. Conclusion

Laser annealing improves the functional properties of super elastic NiTi wire by surface treatment. The phase transformation of martensite to austenite phase during laser heating generates compressive stress on the surface of NiTi wire. Surface annealed of super-elastic, SA wire was achieved by solid state laser beam using both scan speed and applied external stress condition. Both factors have a significant contribution to annealing behavior of the sample. The lower value of the hardness 3.8 GPa of NiTi wire signifies surface annealing of SA sample. The annealed sample shows higher fatigue strength of 6500 cycles in compare to SA sample (3100 cycles). The wires with large residual strain show less fatigue cycle and vice versa.

5. References

[1] Jani J M, Leary L, Subie A, Gibson M.A, 2014 A review of shape memory alloy research, applications and opportunities, *Materials and Design* pp 1078-1113.

[2] Otsuka K and Wayman C. M. 1998 *Shape memory materials, Cambridge: Cambridge university press.*

[3] Otsuka K, Ren X 2005 *Physical metallurgy of Ti-Ni-based shape memory alloys*, Progress in Materials Science pp 511-678.

[4] Tyc O, Plich J, Sittner P 2016, *Fatigue of superelastic NiTi wires with different plateau strain*, Procedia Structural Integrity pp 1489-1496.

[5] Malard B, Plich J, Sittner P, Gartnerova V, Delville R, Schryvers D, Curfs C 2009 *Microstructure and functional property changes in thin NiTi wires heat treated by electric current: high energy X-ray and TEM investigations*, Functional Materials Letters pp 45-54.

[6] Biffi C A and Tuissi A 2017 *Nitinol laser cutting: microstructure and functional properties of femtosecond and continuous wave laser processing*, Smart materials and structures p.035006.

[7] Wang X, Bellouard Y, Vlassak J.J 2005 *Laser annealing of amorphous NiTi shape memory alloy thin films to locally induce shape memory properties*, Acta Materialia pp 4955-4961.

[8] Plich J, Sittner P, Heller L, Delville R, Malard B, Curfs C 2010 Proc. Of SMST.

[9] Plich J, Sittner P, Patent application US20120018413 A1, PCT/CZ2010/000058.

[10] Malard B, Plich J, Sittner P, Delville R, Curfs C 2011 Acta Materialia, pp. 1542-1556.

Acknowledgments

We kindly acknowledge the financial support of this work provided by the Czech Science Foundation by project No. 16-20264S.

IOP Conf. Series: Materials Science and Engineering **362** (2018) 012008 doi:10.1088/1757-899X/362/1/012008

High-temperature Friction and Wear Resistance of Ni-Co-SiC Composite Coatings

Fang Guo, Wan-chang Sun*, Zong-wei Jia, Xiao-jia Liu, Ya-ru Dong

College of Materials Science and Engineering, Xi'an University of Science and Technology, Xi'an, Shaanxi 710054, P.R.China

*Corresponding author. Prof.; sunwanchang@tsinghua.org.cn (Wan-chang Sun); Tel: +86-29-85587373

Abstract: Ni-Co alloy and SiC micro-particles were co-deposited on 45 steel by electrodeposition for high temperature performance. The high temperature tribological characteristics were studied by use of a ball-on-disk method. The micrographs and phase structure of the Ni-Co-SiC composite coatings after high-temperature friction were observed by using a field emission scanning electron microscope(FESEM). The results reveal that the Ni-Co-SiC composite coating presents better wear resistance and lower friction coefficient at high temperature in comparison with that of Ni-Co coating and 45 steel substrate. The embedded SiC particles could strengthen the alloy coating by dispersion strengthening effect and changing the friction mechanism from adhesive wear to abrasive wear.

1. Introduction

In various engineering applications, the long term performance under high-temperature, high-pressure and high-speed conditions in aggressive solution or atmosphere are crucial to mechanical products and parts, made of carbon steels. For example, major failure for hot forging process is originated from high temperature wear, and proportion of which can reach 70-80% [1,2]. Moreover, the majority of engines and cutting tools are also damaged due to the reason mentioned above. In order to improve the wear resistance, corrosion resistance and high temperature oxidation resistance of carbon steels, surface strengthening techniques are needed, of which preparing high-temperature coating is a good choice. Fine particles reinforced metal matrix composite coatings, possessing favorable comprehensive performance, are made up of matrix metal and dispersed ceramic particles such as ZrO_2 [3], B_4C [4] ,Si_3N_4 [5], WC [6] ,Al_2O_3 [3,5,7] ,CeO_2 [8],and SiC particles [5,9] ,among which SiC particles is commonly used as the second ceramic phase owing to its high hardness, low thermal conductivity, relatively high thermal expansion coefficient and high temperature oxidation resistance. On the other hand, Ni-Co alloys present excellent thermal, physical and mechanical properties than pure Ni [10,11]. Therefore, the incorporation of different particles within Ni-Co matrix, achieved by electrodeposition, could lead to a new type of composite coatings that present particular chemical and physical properties [7-9]. Currently, most research works focus on improving the micro-indentation hardness, corrosion resistance, wear resistance, and electro catalytic activity of Ni-Co composite coatings [7-9, 12]. However, there is little report about their high-temperature tribological behavior.

In this work, SiC micro-particles reinforced Ni-Co matrix composite coating which can overcome lots of restrictions under harsh work conditions was prepared by electrodeposition. The high temperature tribological properties of Ni-Co-SiC composite coatings were investigated. For comparison, the high temperature tribological properties of Ni-Co alloy coatings on 45 steel substrates were also investigated.

2. Experimental procedures

The chemical composition and operating condition of electrodeposition of Ni-Co-SiC coatings were collected in Table 1 [13]. The wear resistance of the composite coatings was tested under dry sliding conditions in air at room temperature by the ball-on-disk method, using a high-temperature friction and wear testing machine (HT-1000, Lanzhou Zhongkai LTD., China). The test was carried out with a track radius of 5 mm at 0.3 m/s, using a bearing Si_3N_4 ball (6mm in diameter, ≥HV1300 in micro-indentation hardness) under a load of 5 N for 10 min. The friction coefficient and sliding time were recorded automatically during the test. Each friction pair was cleaned by ultrasonic in acetone before and after each test. The weight loss of the coatings was measured by weighing the samples before and after each wear test. The test conditions were selected based on lots of experiments, according to relevant literature [14-15] . At least five samples of each set of materials were tested for each data, so as to avoid the fluctuations in the data.

The micrographs and phase structure of the Ni-Co-SiC composite coatings after high-temperature friction were observed by using a field emission scanning electron microscope (FESEM, JSM-6390A, Japan).

3. Results and discussion

Fig.1shows the weight loss of 45 steel, Ni-Co coating and Ni-Co-SiC composite coating after 15 min friction. It can be seen that the wear rate decreases from 25.1×10^{-3} mg/m for uncoated substrate to 22.3×10^{-3} mg/m for Ni-Co coating and to 12.6×10^{-3} mg/m for Ni-Co-SiC composite coating. It was clearly demonstrated that Ni-Co-SiC composite coating has the best wear resistance in Fig.1. Furthermore, SiC particles could strengthen the Ni-Co alloy via dispersion strengthening effect [7]. In the meantime, Ni-Co alloy and SiC present different coefficient of thermal expansion, thus the stress between Ni-Co and SiC phases is large enough to form microcracks at high temperature and high pressure. Consequently, lots of tiny particles are easily dug out during the test, whereas, the friction mechanism was changed from adhesive wear to abrasive wear for the existence of many free particles [9, 12]. The weight loss of Ni-Co-SiC composite coating is small, thus exhibiting better high-temperature wear resistance than that of Ni-Co alloy coating, because abrasive wear will not lead to large block debris as adhesive wear.

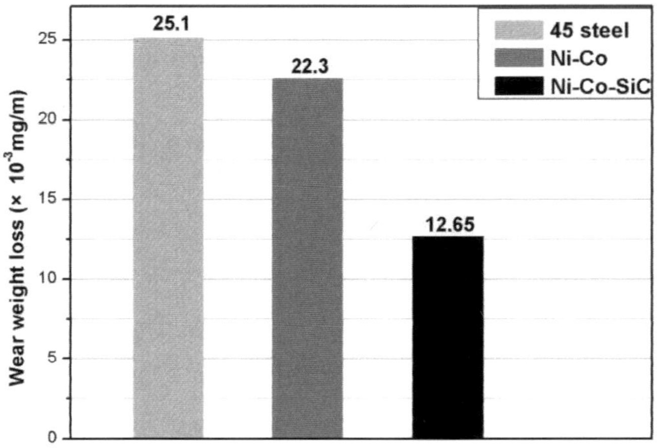

Fig.1. Wear weight loss of 45 steel, Ni-Co alloy coating and Ni-Co-SiC composite coating at 873 K for 15 min.

Fig.2 illustrates the correlation between friction coefficient and sliding time for 45 steel substrate, Ni-Co alloy coating and Ni-Co-SiC composite coating at 873 K. It reveals that the friction coefficient of Ni-Co-SiC composite coating is lower than that of Ni-Co coating, and much lower than that of 45 steel substrate, and which is attributed to different friction mechanisms. For Ni-Co coating, sliding friction is the main form of friction, while the friction mechanism of Ni-Co-SiC composite coating consists of adhesive friction and abrasive attrition. Under high temperature and pressure, the friction mechanism changed and the frictional force reduced greatly when a large number of SiC particles and oxidation scraps were dug out from the coating surface, showing a lower friction coefficient for Ni-Co-SiC composite coating.

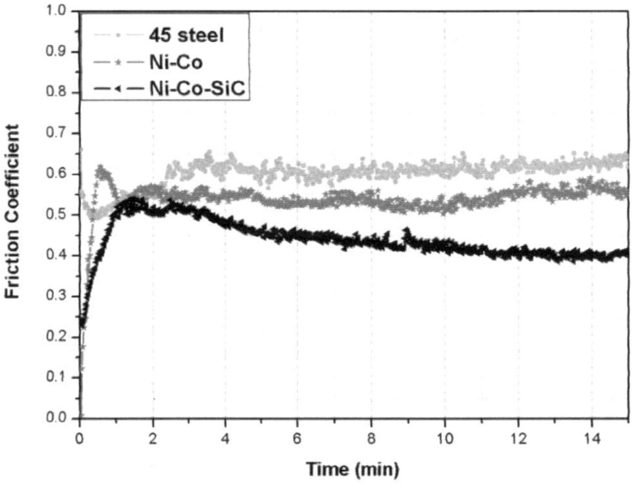

IOP Conf. Series: Materials Science and Engineering **362** (2018) 012008 doi:10.1088/1757-899X/362/1/012008

Fig.2. Friction coefficient of 45 steel, Ni-Co alloy coating and Ni-Co-SiC composite coating at 873 K.

Fig.3 shows SEM morphologies of the worn traces on the surface of 45 steel substrate, Ni-Co and Ni-Co-SiC coatings. Fig.3(a-b) present the morphologies of a typical worn surface of 45 steel substrate at 873 K. Fig.3(a) shows that it has a wide and dark furrow in the sliding direction, which may be formed by plowing and adhesive effect when the sample surface contacts with Si_3N_4 ball. The furrow indicates a higher wear loss. Arrows in Fig.3 (b) show the oxides formed under high temperature. The brittle oxide layer cannot effectively prevent the sliding cut from the contact surface of counter-part. Fig.3(c-d) present the wear traces of Ni-Co coating. Arrows in Fig.3 (d) demonstrate that there are many cracks along the sliding tracks of 45 degree, and thus the integrity of Ni-Co coating is poor. The darker areas which are directed by the arrows are high temperature oxides. The grooves formed on the surface along the sliding direction are most likely as the result of the fragments produced by friction which will cause welding phenomenon and increase the friction force and friction coefficient due to

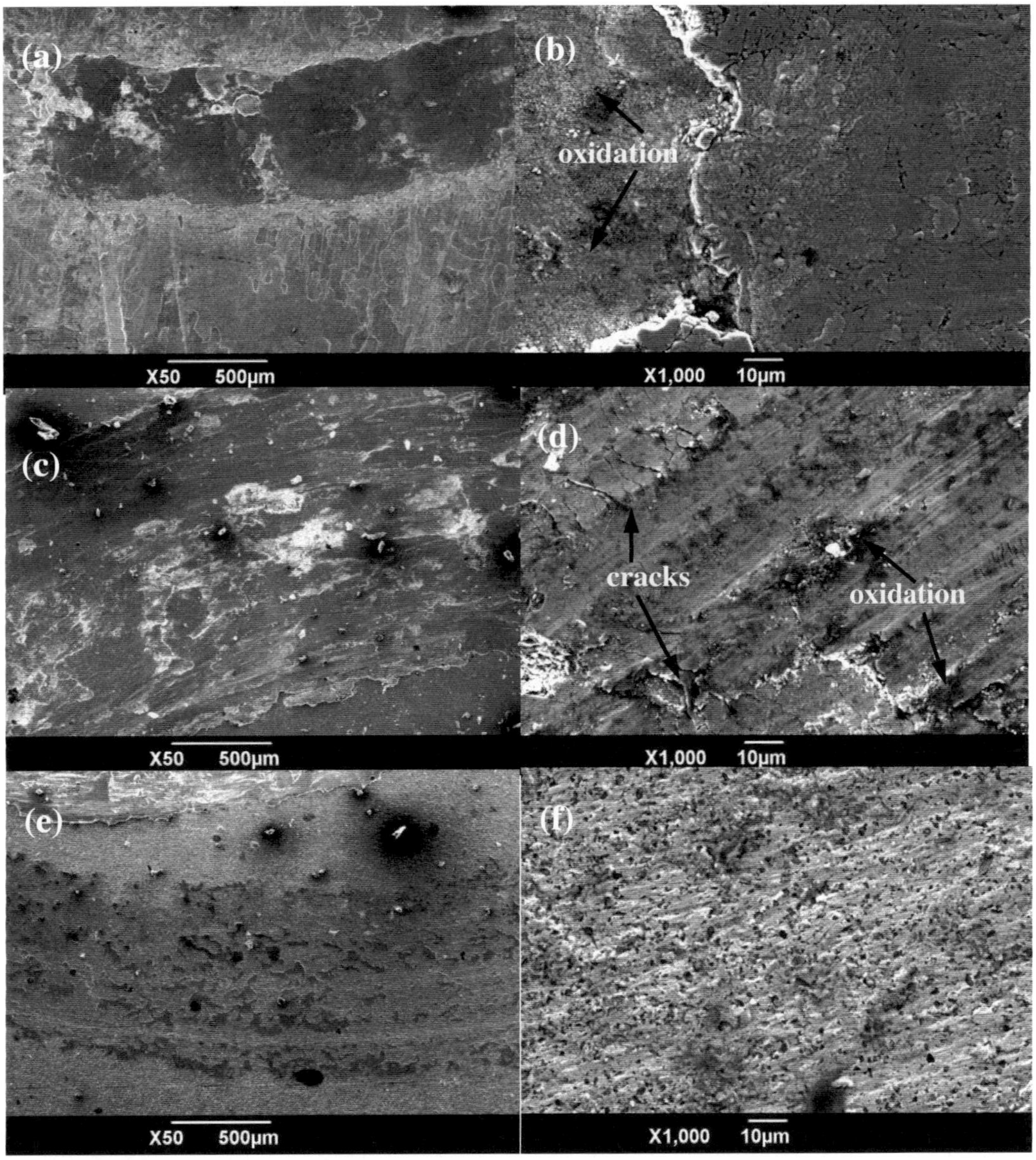

Fig.3. Worn surface morphologies of (a-b) substrate, (c-d) Ni-Co coating and (e-f) Ni-Co-SiC composite coating at 873 K

the presence of the fragments between specimen surface and the pin. Fig.3(e-f) present the worn morphologies of Ni-Co-SiC composite coating, showing that the surface of the composite coating is smooth and continuous, which can be attributed to the effect of embedded SiC particles. The embedded SiC particles carry the load and prevent the matrix alloy from friction by dispersive strengthening effects. And during the wear test, when a large number of ultra-fine SiC particles were dug out by the counter body from the surface of the coating, the friction mechanism was transformed from pure sliding to partly rolling as discussed above, which would reduce the friction force. The friction coefficient of Ni-Co-SiC composite coating is lower than that of Ni-Co coating, confirmed by Fig.2. Because Ni-Co-SiC composite coating possesses the best oxidation resistance, the surface of the composite coating is only slightly oxidized compared with Ni-Co coating and 45 steel, thus reducing the damage of Ni-Co-SiC composite coating, and showing an excellent wear resistance.

4. Conclusions

Ni-Co-SiC composite coatings were prepared by direct current electrodeposition. The embedded SiC particles could strengthen the alloy coating by dispersion strengthening effect, changing the friction mechanism from adhesive wear to abrasive wear. Lower wear loss rate and friction coefficient are presented in Ni-Co-SiC composite coating in comparison with that of Ni-Co coating and 45 steel substrate, indicating excellent wear resistance under high temperature.

5. References

[1] Jiang X, Liu W, Dong S Y, et al. High temperature tribology behaviors of brush plated Ni–W–Co/SiC composite coating[J]. Surface & Coatings Technology, 2005, 194(1):10-15.

[2] Li Z M, Qian, S Q et al. High-temperature tribological properties of Ni-P alloy coatings deposited by electro-brush plating[J].2011, 30(6):669-675.

[3] Gao J, He Y, Wang D. Fabrication and high temperature oxidation resistance of ZrO_2/Al_2O_3, micro-laminated coatings on stainless steel[J]. Materials Chemistry & Physics, 2010, 123(2–3):731-736.

[4] Araghi A, Paydar M H. Wear and corrosion characteristics of electroless Ni-W-P-BC and Ni–P–BC coatings[J]. Tribology - Materials, Surfaces & Interfaces, 2014, 8(3):146-153.

[5] Bai J, Song L, Wang J, et al. Study on PPENK/SiC/Si_3N_4 Heat Resistant Anti-Wearing Nano-Composite Coating[J]. Paint & Coatings Industry, 2013.

[6] Rezaei-Sameti M, Nadali S, Falahatpisheh A, et al. The effects of sodium dodecyl sulfate and sodium saccharin on morphology, hardness and wear behavior of Cr-WC nano-composite coatings[J]. Solid State Communications, 2013, 159(159):18-21.

[7] Wu G, Li N, Zhou, D R, Mitsuo, K, "Electrodeposited Co-Ni-Al_2O_3 composite coatings." Surf. Coat. Technol., 176 157-162 (2004).

[8] Meenu Srivastava, V.K. William Grips, K.S. Rajam. Electrodeposition of Ni–Co composites containing nano-CeO_2, and their structure, properties[J]. Applied Surface Science, 2010, 257(3):717-722.

[9] Shi L, Sun C, Gao P, et al. Mechanical properties and wear and corrosion resistance of electrodeposited Ni–Co/SiC nanocomposite coating[J]. Applied Surface Science, 2006,252(10): 3591-3599.

[10] Srivastava M, Grips V K W, Rajam K S. Structure and properties of electrodeposited Ni–Co–YZA composite coatings[J]. Journal of Applied Electrochemistry, 2008, 38(5):669-677.

[11] Sr M S, Grips V K W, Jain A, et al. Influence of SiC particle size on the structure and tribological properties of Ni–Co composites[J]. Surface & Coatings Technology, 2007, 202(2):310-318.

[12] Wu G, Li N, Dai C S, Zhou D R, "Electrochemical preparation and characteristics of Ni-Co-$LaNi_5$ composite coatings as electrode materials for hydrogen evolution." Mater. Chem. Phys. 83 307-311 (2004).

[13] Wang Y, Sun W C, Xu J M, et al. High-temperature Oxidation of Electrodeposited Ni-Co-SiC Composite Coatings[J]. 2017, 281(1):012013.

[14] Borkar T, Harimkar S P. Effect of electrodeposition conditions and reinforcement content on microstructure and tribological properties of nickel composite coatings[J]. Surface & Coatings Technology, 2011, 205(17):4124-4134.

[15] Zhang, F, Yan, M, "Microstructure and Wear Resistance of in situ Formed Duplex Coating Fabricated by Plasma Nitriding Ti coated 2024 Al Alloy." J. Mater. Sci. Technol. 30 1278-1283(2014).

IOP Conf. Series: Materials Science and Engineering **362** (2018) 012009 doi:10.1088/1757-899X/362/1/012009

Numerical Study of Mechanical Response of Pure Titanium during Shot Peening

Y M Wang, J P Cheng, H P Yang and C H Zhang

School of Metallurgical Engineering, Xi'an University of Architecture and Technology, Xi'an 710055, China
E-mail: ymwang@xauat.edu.cn

Abstract. Mechanical response of pure titanium impacted by a steel ball was simulated using finite element method to investigate stress and strain evolution during shot peening. It is indicated that biaxial residual stress was obtained in the surface layer while in the interior triaxial residual stress existed because the S33 was comparable to S11 and S22. With decreasing the depth from the top surface, the stress was higher during impacting, but the stress relief extent became more significant when the ball rebounded. Therefore the maximum residual stress was formed in the subsurface layer with depth of 130 μm. As for the residual strain, it is shown that the maximum residual strain LE33 was obtained at the depth of 60 μm corresponding to the maximum shear stress during impacting.

1. Introduction

Shot peening is a cold working process used to obtain compressive residual stress and strain hardening. Mechanical properties, such as fatigue and wear resistance, can be enhanced by shot peening [1, 2]. Recently research showed that a nanocrystalline surface layer which is beneficial to mechanical properties could be produced by shot peening [3]. Extensive numerical simulations have been conducted to elaborate the creation of residual stress [4] and the formation of nanoscale microstructure during shot peening [5, 6]. Stress and strain of the target varied significantly with time during shot peening. The evolution led to the residual stress and plastic deformation which correlates with microstructure refinement. Therefore, variations of stress and strain with time are of importance to understand the underlying mechanism of shot peening. However, investigations on the time history of stress and strain are limited [6, 7]. In this study, finite element method (FEM) simulation was carried out to analysis the mechanical response during shot peening in pure titanium which is widely used in many key fields due to their high specific strength and excellent corrosion resistance [8]. Variations of the six components of the stress and strain as the function of time were investigated. Characters of residual stress and residual strain were discussed based on the simulation.

2. Model description

A 3D numerical simulation was carried out by using commercial finite element code ABAQUS/Explicit. A symmetry-cell model with single impacting ball was used to simulate shot peening process [9]. Figure 1 shows the model for single shot impacting. The target with dimension of 2 mm × 2 mm × 4 mm was used. A C3D8R 8-node linear brick element with reduced integral and hourglass control was used for the target. A quarter of the steel ball with a diameter of 0.8 mm was modeled. A C3D4 4-node linear tetrahedron element was used for the ball.

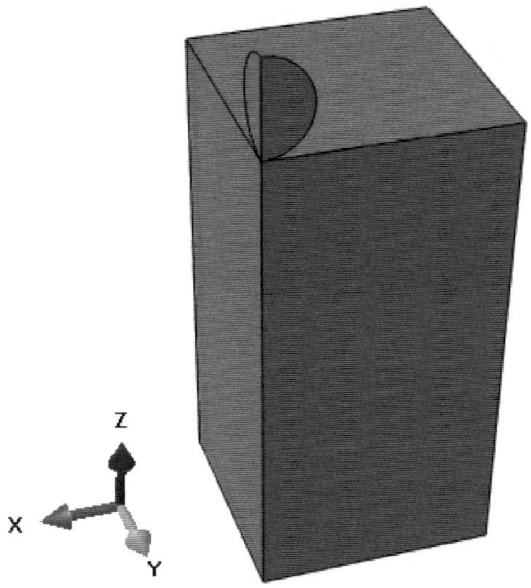

Figure 1. FE model for single shot impacting.

As for boundary conditions, the bottom and two side faces were fixed, and symmetry displacement conditions were applied on the two faces which were in the impacted area. A general contact with an isotropic friction coefficient of 0.1 was used [10].

Plastic deformation behavior of pure titanium was represented by Hollomon's equation as:

$$\sigma = K\varepsilon_p{}^{\mathrm{n}} \qquad (1)$$

where, σ is true stress, ε_p is plastic strain, K is a material constant and n is strain hardening exponent. In this study, K=780.95 MPa and n=0.159 were used. Steel shot was assumed to be elastic during the impacting. The material parameters used in the simulation are summarized in Table 1.

In the simulation, an initial velocity of 80 m/s which is typical for shot peening [11] was imposed on the steel ball. And shot impinging angle 90° was used.

Table 1. Material parameters used in FEM calculation.

	Density (kg/m^3)	Elastic modulus (GPa)	Poisson's ratio
Target	4510	114.8	0.32
Steel shot	7800	200.0	0.26

3. Results and discussion

Figure 2(a) shows the variation of kinetic energy of the system composed of steel ball and target. It can be seen that the energy decreased to the minimum within 1.17×10^{-6} s. Velocity of the ball reached zero at the moment as shown by figure 2(b). After that, the ball begun to move inversely and got a stable velocity of 18.1 m/s at 1.53×10^{-6} s. The kinetic energy was about 1.07×10^{-4} J which was only 6.4% of the initial energy. Fluctuation of the energy shown in figure 2(a) was resulted by the stress wave in the target.

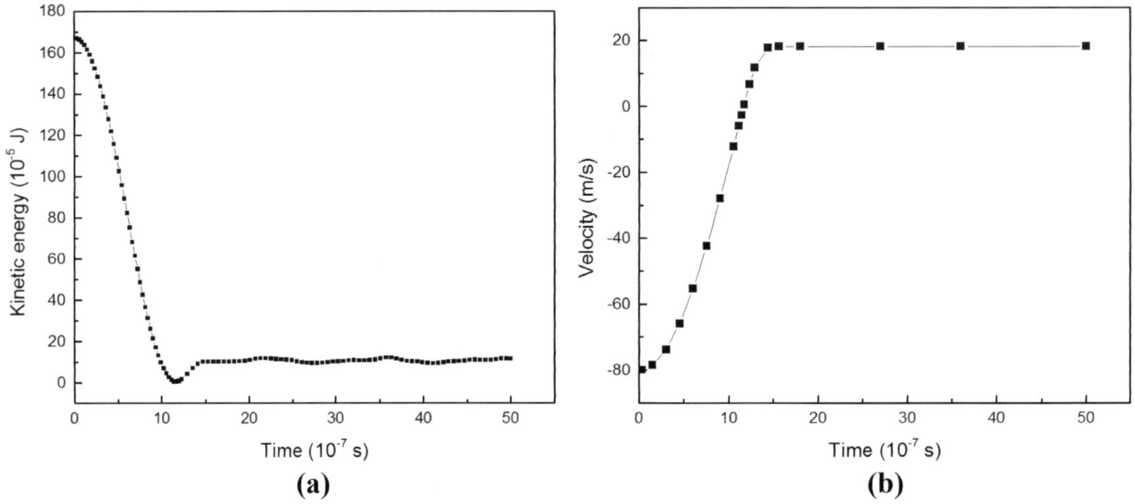

Figure 2. Time history of (a) the total kinetic energy and (b) the velocity of the ball.

Von Mises equivalent stress and plastic equivalent strain (PEEQ) variations with depth from the top surface of the pure titanium at different moments of the shot peening were illustrated by figure 3(a) and (b) respectively. It can be found that the stress level increased with time and reached the highest level at 1.17×10^{-6} s. After that the stress level decreased. At different moments, the maximum stresses occurred in the subsurface. At 1.17×10^{-6} s when the velocity reached zero, the depth at which the maximum stress occurred was about 60 μm. After the ball rebounded from the target, the depth was about 130 μm. The variation curves of the of Von Mises equivalent stress with time are not smooth at some depths because the change of stress sign. Similar to the Von Mises equivalent stress, the maximum of PEEQ also occurred in the subsurface. At the moment when the velocity was zero, the depth at which the maximum PEEQ 0.30 existed was about 60 μm. After that, the depth almost kept stable.

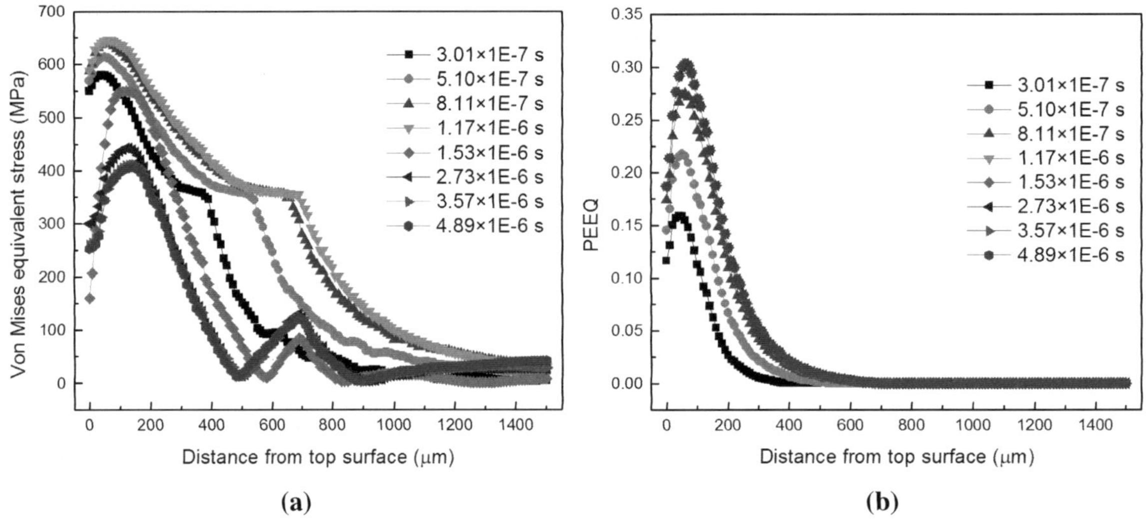

Figure 3. Distribution of (a) the Von Mises equivalent stress and (b) the PEEQ at different moments.

Variation of the stress components with time at different depths from the top surface of the pure titanium are shown in figure 4. Clearly, shear stresses were almost zero for the titanium at different depths and different moments. During impacting the normal stresses increased with time and the magnitudes of S33 were larger than S11 and S22 obviously. When the ball rebounded, the normal stresses decreased quickly especially the S33. At last, S33 nearly reached zero at the top surface.

IOP Conf. Series: Materials Science and Engineering **362** (2018) 012009 doi:10.1088/1757-899X/362/1/012009

Biaxial residual stress state was obtained in the surface layer. The residual S33 increased with the depth, which is shown by figure 4. It was found that at the depth of 400 μm, the residual S11 and S22 were about 309 MPa, while the S33 was about 222 MPa. Therefore, triaxial residual stress state was obtained in the interior because the S33 could not be neglected. Though the stress level was not high at the depth of 130 μm during impacting compared with the stress level in the top surface and at the depth of 60 μm, the residual stress was the highest because the stress relief was weakened with the increase of depth. In addition, fluctuation of the normal stresses can also be found because of the stress wave.

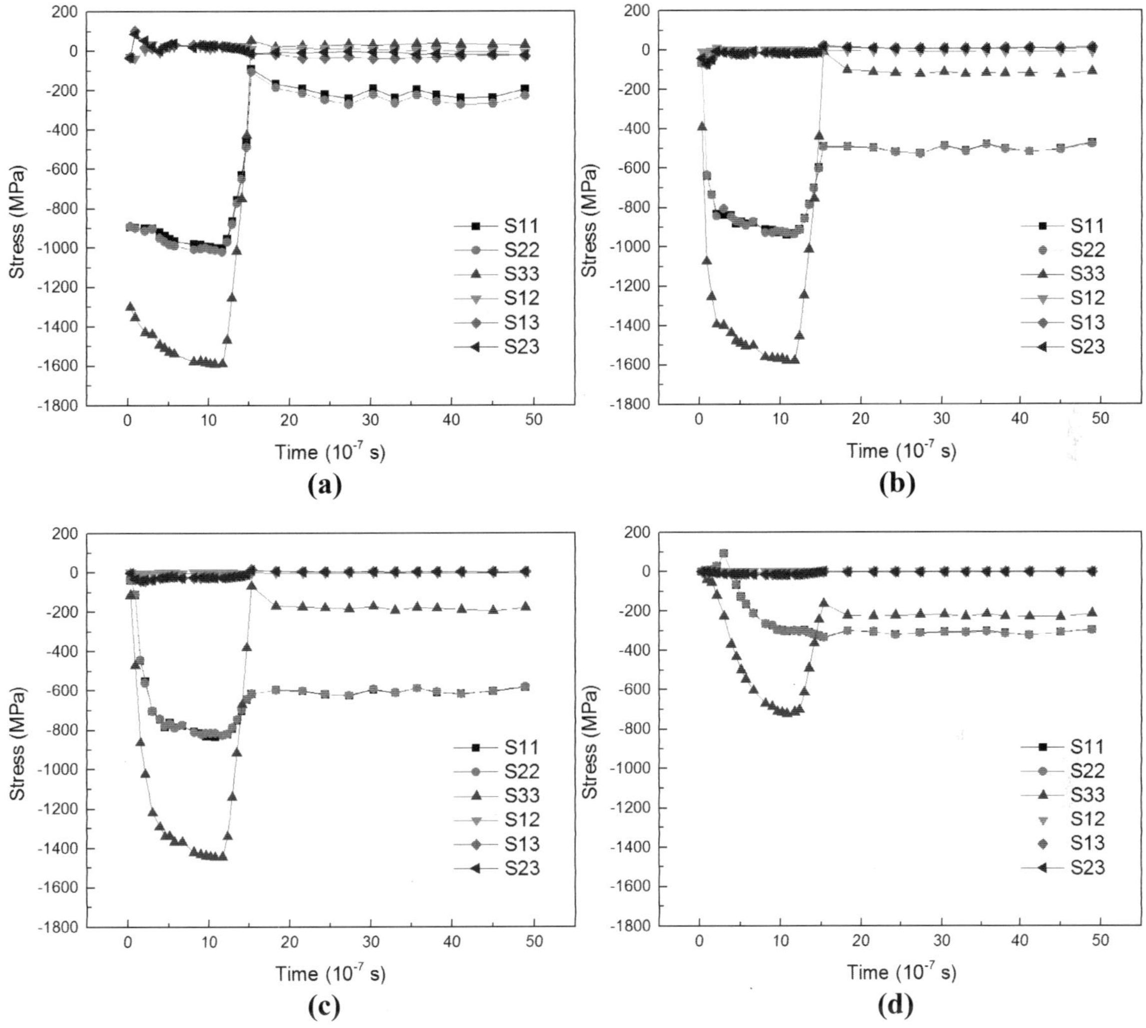

Figure 4. Time histories of stress components (a) in the top surface and at the depth of (b) 60 μm , (c) 130 μm and (d) 400 μm from the top surface.

Corresponding to the stress components in figure 4, variation of strain components were shown in figure 5. It is found that LE11 was equal to LE22, and the magnitude of LE11 or LE22 was half of LE33. LE12 was almost zero. LE13 and LE23 were equal. When the ball rebounded, strains relaxed slightly. It can be found the maximum strain occurred at the depth of 60 μm, which was agreement with the PEEQ distribution. Residual strain at depth of 130 μm where maximum residual stress occurred was obvious smaller compared with the residual strain at the depth of 60 μm. This means that the variation of residual strain with depth was different from the variation of residual stress with depth. If the shear stresses were not considered as shown in figure 4, the normal stress S33, S11 and S22 could be treated as principle stresses. In this case, it is easy to find that the maximum shear stress existed at the depth of 60 μm. Because the plastic deformation was induced by shear stress, the maximum plastic deformation was reasonably obtained at the depth corresponding to the maximum shear stress.

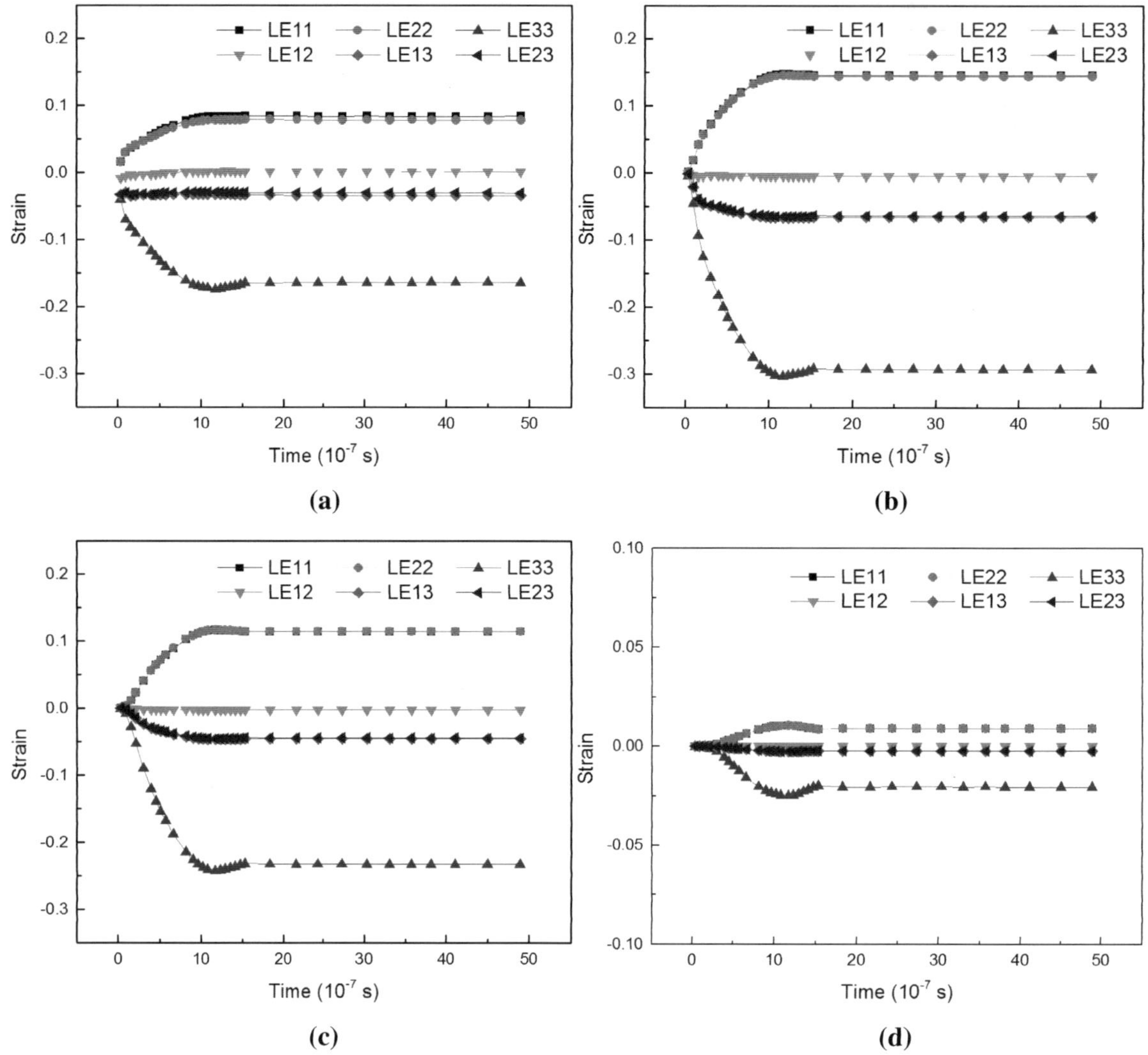

Figure 5. Time histories of strain components (a) in the top surface and at the depth of (b) 60 μm , (c) 130 μm and (d) 400 μm from the top surface.

4. Conclusions

In this study, shot peening was modeled by single shot impacting using FEM simulation. Based on the results and discussion, following conclusions can be drawn:

(1) Shear stress was nearly zero. The normal stress S33 along the impacting direction was the most significant during impacting. All the normal stress components especially S33 released quickly when the ball rebounded, therefore biaxial residual stress was obtained in the surface layer. In the interior triaxial residual stress existed because the S33 was comparable to S11 and S22.

(2) The residual stress is influenced by the stress state during impacting and the stress relief when the ball rebounded. With decreasing the depth from the top surface, the stress was higher during impacting, but the stress relief extent became more significant. Therefore the maximum residual stress was formed in the subsurface layer.

(3) Strain component LE12 was nearly zero. LE13 and LE23 were equal and they were small compared with normal strain components. LE11 was equal to LE22, and the magnitude of them was half of LE33.

(4) The correlation between residual stress and depth is different from that between residual strain and depth. The maximum residual stress was obtained at the depth of 130 μm while the maximum residual strain was at the depth of 60 μm corresponding to the maximum shear stress.

Acknowledgments
Financial support from Science and Technology Department of Shaanxi Province (2017GY-115) and Education Department of Shaanxi Province (16JK1466) are gratefully acknowledged.

References

[1] Soady K A 2013 *Mater. Sci. Technol.* **29** 637-51
[2] Mitrovic S, Adamovic D, Zivic F, Dzunic D and Pantic M 2014 *Appl. Surf. Sci.***290** 223-32
[3] Bagheri S and Guagliano M 2009 *Surf. Eng.* **25** 3-14
[4] Zimmermann M, Klemenz M and Schulze V 2010 *Int. J. Comput. Mater. Sci. Surf. Eng.* **3** 289-310
[5] Bagherifard S, Ghelichi R and Guagliano M 2010 *Surf. Coat. Technol.* **204** 4081-90
[6] Dai K and Shaw L 2007 *Mater. Sci. Eng.* **A463** 46-53
[7] Meguid S A, Shagal G and Stranart J C 1999 *J. Mater. Process. Technol.***92-93** 401-4
[8] Lütjering G and Williams J C 2007 *Titanium* (Berlin: Springer-Verlag)
[9] Hong T, Ooi J Y and Shaw B A 2008 *Adv. Eng. Software.* **39** 743-56
[10] Shivpuri R, Cheng X and Mao Y 2009 *Mater. Des.* **30** 3112-20
[11] Evans R W 2002 *Mater. Sci. Technol.* **18** 831-9

IOP Conf. Series: Materials Science and Engineering **362** (2018) 012010 doi:10.1088/1757-899X/362/1/012010

The Deformation Mechanism of Fatigue Behaviour in a N36 Zirconium Alloy

Yingzhu Wang

Huanggang Middle School Guangzhou Branch
+86 18566395386
1476456993@qq.com

Abstract. Zirconium alloys are widely used as claddings in nuclear reactor. A N36 zirconium alloy has been deformed into a sheet with highly texture according to the result of electron back scatter diffraction test. Then this N36 zirconium alloy sheet has been cut into small beam samples with 12 x 3 x 3 mm^3 in size. In this experiment, a three-point bending test was carried out to investigate the fatigue behaviour of N36 zirconium alloy. Cyclic loadings were applied on the top middle of the beam samples. The region of interest (ROI) is located at the middle bottom of the front face of the beam sample where slip band was observed in deformed beam sample due to strain concentration by using scanning electron microscopy. Twinning also plays an important role to accommodate the plastic deformation of N36 zirconium alloy in fatigue, which displays competition with slip.

1. Introduction

Research into nuclear power began over 65 years ago, however, incidents such as the accident in Pennsylvania in 1979 and the disaster at Chernobyl in 1986 triggered increasing concern about the safety in nuclear industrial. There has been an increased use of zirconium alloys as fuel cladding within the nuclear industry since the late 1950s owing to a combination of their good corrosion resistance at operating temperatures and low neutron absorption. The cladding also protects the radioactive fuel from the surrounding environment whilst providing a good heat transfer surface. It is the value of the absorption cross-section that is most relevant for materials with nuclear uses.

Zirconium alloys used within the nuclear reactor are normally considered structural components, and their performance during operation is dependent on the deformation mechanisms that occur during operation and those that give rise to the final microstructure of the alloy as a result of the processing route. Deformation during operation normally involves elastic deformation, during which the material will return to its original shape, unless the material undergoes a stress which will lead to failure, as might occur during a loss of coolant accident (LOCA). During processing, the alloy normally undergoes a significant amount of plastic deformation which gives rise to the alloy's final microstructure and the mechanical properties it will exhibit during service. The deformation behaviour of zirconium alloys during processing and failure scenarios will be discussed in this section. Work done by Akhtar et al. [1] investigated the slip system operative for various crystallographic orientations and how these, along with a number of other conditions, effect the work hardening parameters. During this study, single crystal specimens were found to be most suitable when investigating deformation mechanisms under stress as the complex effects of grain boundaries do not have to be taken into consideration.

Twinning plays a significant role in the deformation of zirconium and zirconium alloys. Crystallographic twinning occurs when two separate crystals share some of the same crystal lattice points in a symmetrical manner. The preference type of twinning including the typical

IOP Conf. Series: Materials Science and Engineering **362** (2018) 012010 doi:10.1088/1757-899X/362/1/012010

shears and reorientations in zirconium has been concluded in previous research [2, 3] as shown in Fig. 1: $\{10\bar{1}2\}$ $<\bar{1}011>$ twinning and $\{11\bar{2}1\}$ $<\bar{1}\bar{1}26>$ twinning are activated when zirconium is under tensile stress along the direction of the c axis; when zirconium is under compression along the <c> direction $\{11\bar{2}2\}$ $<\bar{1}\bar{1}23>$ twinning and $\{1011\}$ $<1012>$ twinning are observed at room temperature and elevated temperatures, respectively.

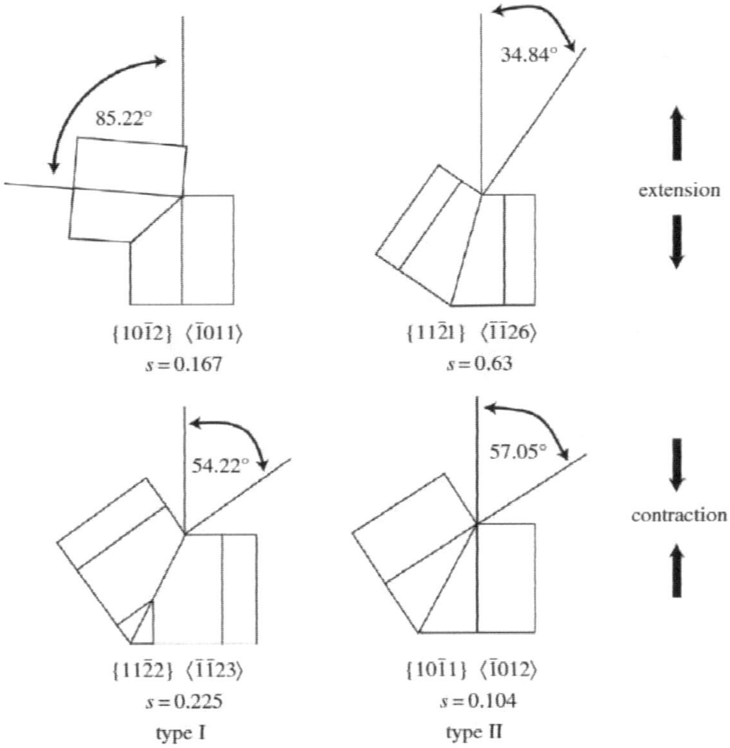

Figure 1. Twinning type in zirconium [2, 3].

Twinning causes a region within a parent grain which has a shear strain. This shear strain contributes to a crystallographic reorientation with respect to the parent crystal [3]. Twinning has a competition with slip during the deformation because twinning not only produces strains in the uniaxial extension or compression of the c axis, but also accommodates shears in the basal plane, and in the latter case twinning is the dominated mode [4]. Twinning contributes to the reorientatioin of the grain like slip, and it forms subgrains with different orientation compared with the matrix [5]. Kaschner et al [6] reported the effect of twinning on the hardening response in zirconium that twins influence the reload behaviour by either reorienting material for easy slip or by providing barriers to slip.

In this study, a three-point bending test was designed to investigate the deformation of N36 zirconium alloy. A finite element modelling was built up to estimate the deformation of N36 zirconium alloy. Additionally, high resolution digital image correlation (DIC) and electron backscatter diffraction (EBSD) technology were used to acquire the strain distribution and grain orientation in N36 zirconium alloy. These microstructure knowledge contributes to a deeper understanding of the deformation mechanism in zirconium. The relationship between slip and twinning has been discussed.

2. Methods and materials

2.1. Chemical composition of N36 alloy

The as-received material is a cold-rolled sheet of N36 alloy with the thickness at 3 mm. The chemical composition of N36 alloy is given in Table 1.

Table 1-1 Chemical composition of N36 alloy

Element	Sn	Fe	Cr	O
Percentage (%)	1.45	0.27	0.10	0.09

2.2. Three-point bending test

The as-received sheet of N36 alloy has been cut into beam samples with 12 × 3 × 3 mm^3 in size as shown in Fig. 2(a). These samples were then deformed in a three-point bending test on a Zwick machine by applying cyclic loading. The schematic diagram of the three-point bending is shown in Fig. 2(b). The force-controlled cyclic loading was applied on the top middle of beam. Two cylinder pins were fixed to support the beam in the bending test, which make the beam not free at vertical direction. In beam theory, stress is expected to concentrate at middle bottom of beam which is marked in red. Thus, the region of interest (ROI) is focused at this red section. The loading history is plotted in Fig. 2(c).

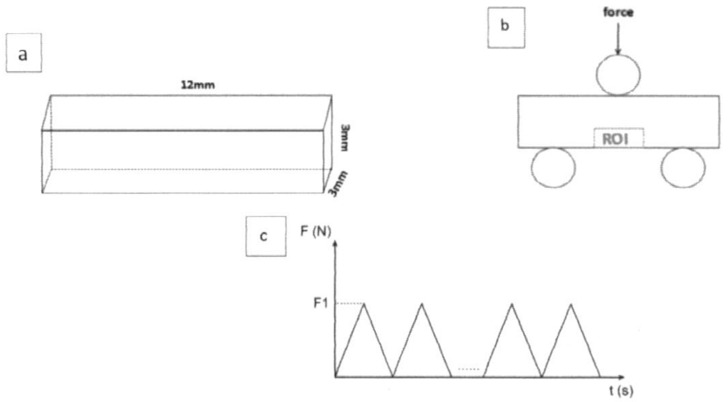

Figure 2. Schematic diagram of three-point bending test (a) three-point bending test setting; (b) the size of beam sample; (c) loading history of cyclic loading

2.3. High-resolution digital image correlation (HR-DIC)

Digital image correlation (DIC) technology was developed to measure the strain distribution in the ROI during the three-point bending. The basic method of DIC is to measure the displacement change by tracking the characteristic on the surface of the sample. In this experiment 250 nm particles of diamond were sprayed on the sample surface forming a uniform distribution of spot coating as shown in Fig.3. After each loading increment, the beam sample was removed from three-point bending rig to Sigma 300 Scanning Electron Microscopy (SEM). The morphology of the speckle coating was observed by SEM. Based on a cross-correlation method by an in-house coding the strain distribution of deformation in beam was obtained.

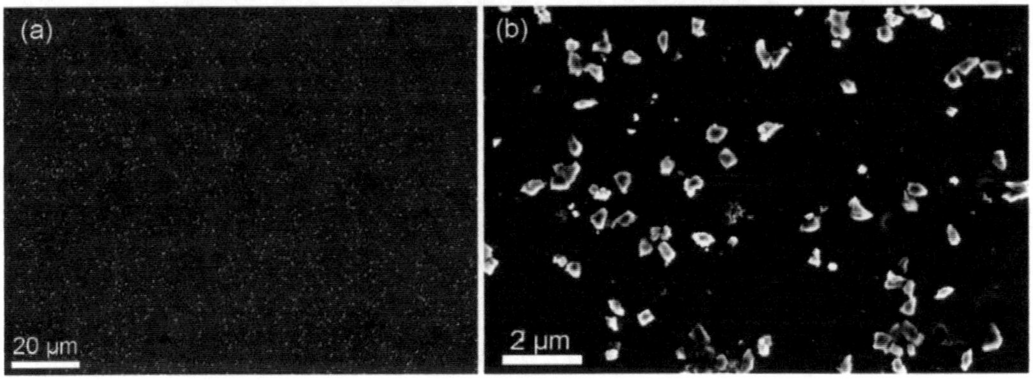

Figure 3. Particles sprayed on the surface of the sample for DIC measurement. (a) low magnification image; (b) high magnification image

2.4. EBSD and High-resolution EBSD

Electron backscatter diffraction (EBSD) is a technique for automatically obtain information about crystallographic orientation from samples at each individual scanning point in a SEM. The Zeiss Auriga-40 SEM and Bruker EBSD detector was used to observe the free surface of samples in this experiment. EBSD is not only used to measure the texture, ODF and MODF in materials, but also can be used to analysis the phase composition by checking Kikuchi pattern.

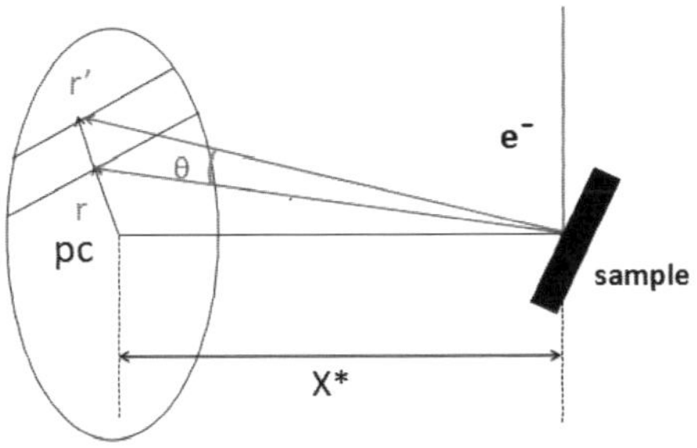

Figure 4. Schematic of EBSD showing the band width relation

Fig. 4 shows the Fundamental of EBSD. The samples in the EBSD test were tilted by 70 degrees from the horizontal plane. The incident electrons that satisfy the Prague law cast the crystal face onto the screen forming the Kikuchi band. The width of the bands is a direct function of the d-spacing of the diffracting plane through Bragg's law:

$$\lambda = 2d_{hkl}sin\theta,$$

λ is the wavelength of the incident wave, d_{hkl} is the d-spacing of the reflection plane, and θ is the angle prescribing the width of the corresponding Kikuchi band.

3. Finite element modelling

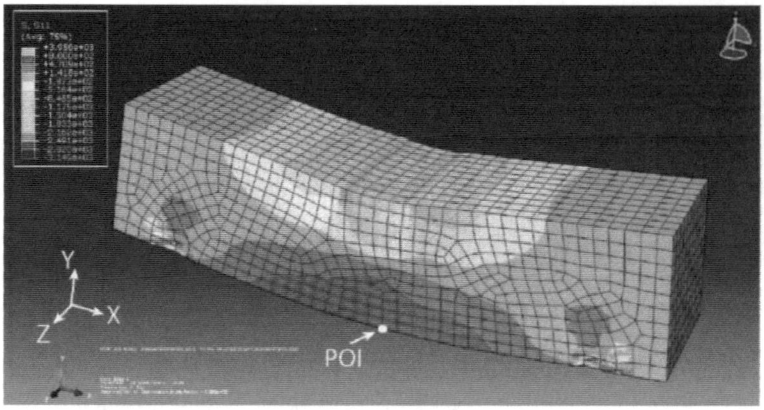

Figure 5. Three-point bending finite element simulation of N36 zirconium alloy beam

Fig 5 shows the simulation of the stress distribution of a zirconium alloy beam during a three-point bending test. This beam was applied by a cyclic loading of a magnitude of 1300 N at the middle top. This beam was placed in a special three-point bending setting where it was supported by two cylinder pins. In this way there has no freedom for the beam to move in the vertical direction but it only can be moved in the horizontal direction. The point of interest (POI) is located at the middle bottom of the beam shown as white point in the figure. From the simulation results it can be seen that the stress at POI is the highest at 800 MPa during deformation compared with the surrounding region. In the area around the POI, the farther away from the POI the smaller the stress value is. The stress also highly concentrated at the area close to the two supporters. However, since the positions of the two supporters are far away from the POI the stress distribution at the POI is not affected.

4. Results and discussion

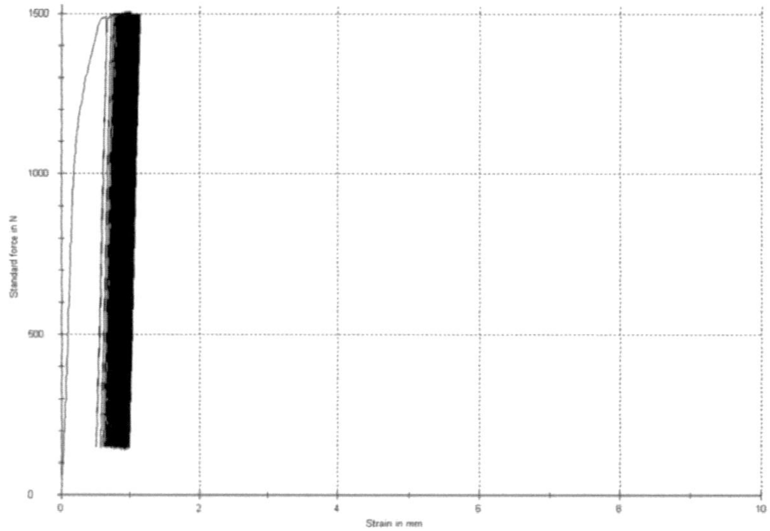

Figure 6. Force versus displacement (strain in mm) curve of cyclic loading test

Fig.6 shows the relationship between applied force (standard force) and displacement (strain in mm) during a cyclic loading. The maximum applied force is 1500N and the minimum one is 150 N with the ratio, R = 0.1. At the first cycle the alloy shows obvious plastic deformation with displacement remains around 0.5 mm after unload. After that, the displacement increases at every cycle. However, the increase value of displacement of every cycle reduces progressively. After 100 cycle, the increment of displacement of every cycle is relatively low but the total accumulation is pretty large which reaches 1 mm at 2000 cycle.

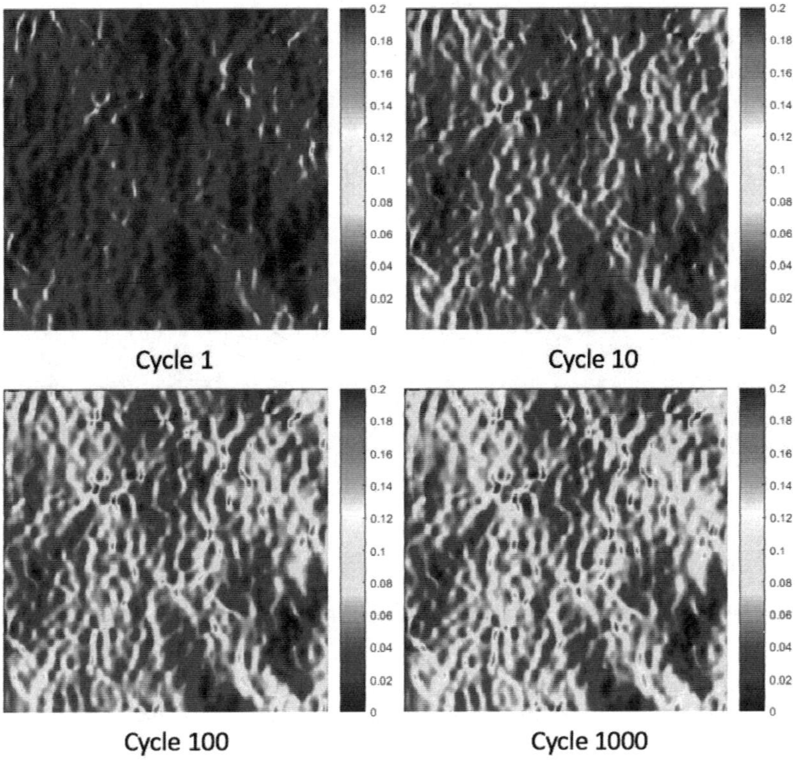

Figure 7. DIC results of three-point bending N36 alloy showing the strain distribution in the area of interest

Fig 7 illustrates the DIC result of alloy during the three-point bending cyclic loading. These DIC results give the total distribution of cycle 1, cycle 10, cycle 100 and cycle 1000 at the region of interest. The strain shows apparently localization in the grain or at the grain boundary. From cycle 1 to cycle 10, the strain localization enhances obviously with the maximum strain reaches 0.2. After the cycle 10, the strain increases continuously but shows not much difference between cycle 100 and cycle 1000.

IOP Conf. Series: Materials Science and Engineering **362** (2018) 012010 doi:10.1088/1757-899X/362/1/012010

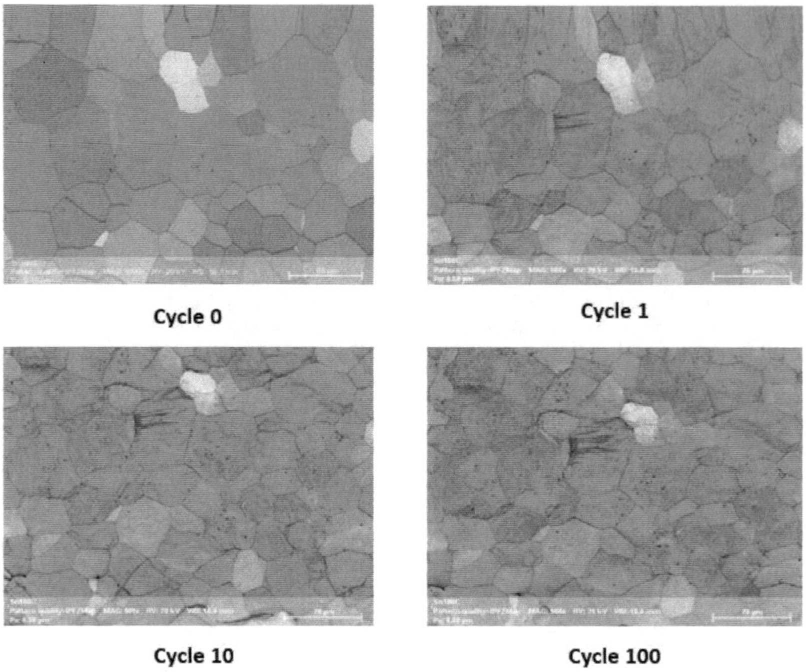

Figure 8. EBSD results of cyclic loading three-point bending test

Fig 8 shows the EBSD results of N36 alloy applied by cyclic loading from cycle 0 to cycle 100. The observation on the surface of electropolished sample gives a satisfactory view on the orientation variation of grains in the region of interests, which indicates a strong basal texture in this sample. The grain size ranges from 10 μm to 50 μm. The deformed grains represent shape variation and lattice distortion, which contributes to the formation of subgrains. In cycle 0, the original situation of the grains shows no twinning or slip band. However, twinning, marked as blue area, appears from cycle 1 and enhances with the increase of cycle number. Meanwhile, only a few slip bands can be seen in the grain but it does not dominate the deformation. It indicates that twinning is necessary for the deformation of this alloy.

5. Conclusion

Three-point bending test on beam samples combining with finite element modelling analysis and experimental method was carried out to investigate the deformation of N36 zirconium alloy. The conclusions below have been obtained:

(1) The modelling results indicates that the stress peaked at the middle bottom of the beam sample.

(2) Strain localization was observed in N36 zirconium alloy during the cyclic loading three-point bending test which indicates the localization of slip band or the existence of twinning.

(3) Twinning dominates the deformation of N36 zirconium alloy applied at three-point bending test displaying competition with slip according the analysis on EBSD results.

6. References

[1] A. Akhtar, Teghtsoo.A, Acta Metallurgica, 19 (1971) 655-&.

[2] E. Tenckhoff, Deformation Mechanisms, Textures, and Anisotropy in Zirconium, 1988.

[3] T.B. Britton, F.P.E. Dunne, A.J. Wilkinson, Proceedings of the Royal Society A: Mathematical, Physical and Engineering Science, 471 (2015) 20140881.

[4] C.N. Tome, R.A. Lebensohn, U.F. Kocks, Acta Metall Mater, 39 (1991) 2667-2680.

[5] B. Clausen, C.N. Tomé, D.W. Brown, S.R. Agnew, Acta Materialia, 56 (2008) 2456-2468.

[6] G.C. Kaschner, C.N. Tomé, I.J. Beyerlein, S.C. Vogel, D.W. Brown, R.J. McCabe, Acta Materialia, 54 (2006) 2887-2896.

IOP Conf. Series: Materials Science and Engineering **362** (2018) 012011 doi:10.1088/1757-899X/362/1/012011

A Methodology for Evaluating the Hygroscopic Behavior of Wood in Adaptive Building Skins using Motion Grammar

Rana El-Dabaa

Department of Architectural Engineering & Environmental Design
Arab Academy for Science, Technology and Maritime Transport
Cairo, Egypt
rana.bahaa@aast.edu

Sherif Abdelmohsen

Department of Architecture
The American University in Cairo
Cairo, Egypt

sherifmorad@aucegypt.edu

Abstract. The challenge in designing kinetic architecture lies in the lack of applying computational design and human computer interaction to successfully design intelligent and interactive interfaces. The use of 'programmable materials' as specifically fabricated composite materials that afford motion upon stimulation is promising for low-cost low-tech systems for kinetic facades in buildings. Despite efforts to develop working prototypes, there has been no clear methodological framework for understanding and controlling the behavior of programmable materials or for using them for such purposes. This paper introduces a methodology for evaluating the motion acquired from programmed material – resulting from the hygroscopic behavior of wood – through 'motion grammar'. Motion grammar typically allows for the explanation of desired motion control in a computationally tractable method. The paper analyzed and evaluated motion parameters related to the hygroscopic properties and behavior of wood, and introduce a framework for tracking and controlling wood as a programmable material for kinetic architecture.

1. Introduction

Interactive architecture generally lies at the intersection between digital technologies and physical spatial experiences [1]. Gu and Wan define it as the intersection of four disciplines: computer science, interaction and experience design, architecture and behavior and social studies [2]. The technology behind interactive architecture is influenced by many disciplines including physics, biology, mechanics and electrical engineering and human computer interaction (HCI). Smart materials are increasingly becoming a significant constituent in such systems, where actuation and sensing properties are currently embedded in everyday objects and surfaces [3].

As per Mark Weiser's observation, "The most profound technologies are those that disappear. They weave themselves into the fabric of everyday life until they are indistinguishable from it." This has encouraged researchers in the fields of pervasive and ubiquitous computing and the Internet of Things to introduce new interfaces, such as Tangible User Interfaces (TUIs). Interactivity in tangible interfaces is driven from physical elements or devices in the physical environment with properties of objects such as appearance or sound. The recent shift from tangible user interfaces to natural and organic user interfaces – and more recently smart material interfaces (SMIs) – has dramatically pushed the limits of interactive architecture, where the seamless interaction between the physical and the digital occurs primarily through material properties [4].

Traditionally, the development of interactive and kinetic systems, specifically for building facades, relied heavily on the implementation of sophisticated hard mechanistic systems. Fox and Kemp argue for the "*End of Mechanisms*"; a paradigm shift in kinetic systems from using mechanical devices to passive biological systems [5]. The use of smart materials was proposed as a part of this argument, due to the close relation between smart material behavior and the mimicking of natural behavior [6].

An inherent property in smart materials is their responsive behavior, where they can change their properties and can exhibit control and reversibility to respond to environmental stimuli. Programmable material is a type of smart material in which this composite material is fabricated and programmed to respond to external stimuli with a specific desired behavior. This phenomenon can be directly applied in interactive and responsive architecture.

This paper attempts to identify a theoretical basis for capturing and controlling the movement of programmable materials – specifically as related to the hygroscopic properties of wood – for further implementation in kinetic and responsive structures. This study explores motion grammar as a method for capturing and analyzing motion in programmable material. Literature in motion grammars is primarily discussed from two viewpoints: linguistics and computer science, and image processing in the robotics domain [7]. Little has been done however to replace mechanical actuators in responsive systems with programmable materials. This requires the definition of a clear methodology to investigate, capture and control the behavior of smart materials. This paper addresses the following questions: How does material fabrication and development affect the realm of responsive and kinetic architecture? How can programmable material behavior be analyzed and captured through Motion Grammars?

2. Methodology

The challenge in applying Motion Grammars as a method to capture the motion of programmable material lies in the inherent properties of that material. This material is analyzed in relation to the various components of robotic systems and series of tasks done. Hygroscopic property of wood is used as a main generator for programming wood motion according to moisture content. A series of experiments are conducted to analyze and understand the behavior and response of wood to moisture content. By comparing and analyzing the motion and process of robotic systems to that of the hygroscopic response of wood, a set of Motion Grammar parameters are deduced.

3. Embedded Material Intelligence

Smart materials are generally classified and defined in different ways. Addington & Schodek differentiate between smart and traditional materials via five main characteristics: transition, selectivity, immediacy, self-actuation and directness. The capabilities of materials with these properties can be divided into property changing capability, energy exchange capability, discrete size/location and reversibility. One of the most significant characteristics of smart materials is the ability to change in size. They are also known to have either reversibility or bi-directionality properties. Reversibility is the ability of a material to exchange its input and output form, while bi-directionality is the ability of a material to change either its input or output energy properties as phase changing materials [8]. Ritter also classified smart materials as those that can reversibly and repeatedly change their properties in response to external stimuli as changes in temperature or light. Semi-smart materials can change their properties once or a limited number of times. Smart and semi smart materials are called functional materials [9]. The term "Stimuli-responsive

IOP Conf. Series: Materials Science and Engineering **362** (2018) 012011 doi:10.1088/1757-899X/362/1/012011

materials" (SRMs) is also used to denote materials that alter their properties according to external stimuli [10].

Another classification that relies on the relation between materials, technologies and environment, divides smart materials into two types; property changing, and energy exchanging materials. Property changing materials are those whose properties change according to an energy input, such as electro chromic glazing. Energy exchanging materials are those that transform the material input energy into another form of energy, such as photovoltaics, which transform solar radiation into electric energy [8].

Shape-shifting materials are classified according to their trigger stimuli into photostrictive materials that are stimulated by light, thermostrictive that are stimulated by the effect of thermal energy, piezoelectric that are stimulated by the effect of pressure and tension, electroactive that are stimulated by electrical energy, magnetostrictive that are stimulated by the effect of magnetic fields, and chemostrictive that are stimulated by chemical energy. According to Ritter's classification, shape shifting materials encompass thermostatic materials and electroactive smart materials [9].

Thermo-bimetals (TB) are another recently applied example of smart materials. They are laminated composite materials composed of a minimum of two metal layers having different thermal expansion coefficients. The layer with lower thermal expansion coefficient is the passive layer, while the active layer is the one having the higher thermal expansion coefficient. By changing the temperature, the composite material changes from a straight to a curved strip with the active layer facing upward and the passive layer facing downward.

This paper focuses on the use of wood as a programmable material due to its hygroscopic properties. Relevant work includes the efforts at the Institute for Computational Design (ICD) at the University of Stuttgart, and the Self-Assembly Lab (SAL) at MIT. Various experiments have been done to maintain, control and enhance the embedded motion in these materials [11]. Hygroscopic design in wood has been introduced to programmable materials due to the ability to program the response of wood specifically in relation to moisture content [12]. Wood is stimulated by humidity in the air and the moisture content inside wood. Figure shows the response of a beech veneer sample exposed to humidity. The continuous shrinking and swelling depends on relative humidity and the moisture content inside the sample.

Wood is selected for its low carbon footprint and embodied energy [13]. Studies have shown different ways in controlling the motion of wood through grain orientation, thickness and layers of lamination [14]. This research builds on these studies and uses Motion Grammars to investigate the type of motion taking place in wood upon increase in moisture content.

Figure 1. Time lapse for wood response during increase in moisture content.

4. Grammars

Formal languages are outlined by grammars that demonstrate methods to control systems and express full system dynamics [15]. Shape grammars, the earliest visual method of design and analysis for generative purposes introduced in the early 1970's, are a group of transformation rules that are applied on a shape to generate a new one [16]. This computational method relies on defining a specific initial shape, and a rule which identifies two parameters; the shape that will be replaced and the method it will be replaced with [17].

Motion Grammar differs, as it is targets the 3D movement of tools or materials; thus leading to generative ways to investigate output designs [18]. Motion Grammar is generally the linguistic system application that gives the control to a robot. Significant difference is found between the language of the system and language of specification in linguistic control methods. The system represents the physical object itself as

IOP Conf. Series: Materials Science and Engineering **362** (2018) 012011 doi:10.1088/1757-899X/362/1/012011

the controller, while the specification indicates the response behavior of the object. Both are located as formal languages, and the specifications are categorized as formal languages [7]. Motion in hybrid systems as humans and robots are described as continuous and distinct elements. Formal languages are formed through groups of strings. Discrete events and actions are illustrated as a series of atomic symbols; that are called strings [15].

The grammar states all the options of change in assets from the morphemes to achieved basic motions. Combining and sorting these motions are the rules that produce composite motions [18]. The grammar exemplifies continuous dynamics through what is known as Context-Free Grammar (CFG); that depends on isolated strings. These strings are used to replace the variables and terminals, thus forming productions. CFG encompasses a set of productions [19]. This paper focuses on motion grammar as a method to analyze the response behavior of programmable material.

5. Motion Parameters

Elements forming a grammar are variables, terminals and productions (rules). Variables are the initial variables, presented in uppercase letters (also called non-terminals). Terminals are represented in alphabets, and productions are strings that form the rules [19].

Introducing grammars as a way to describe design was developed to encompass parametric design, shape grammar and generative fabrication [18]. Lately, studies have been conducted to define motion in relation to formal grammars [17], including any transformation in position or orientation through time. This refers to the process of change, even if the beginning and end points are the same [18]. Other studies attempted to link tangible elements of kinematics like structure and materials with intangible elements like motion, through the pedagogical learning process of "Imitation, Iteration and Improvisation" that focuses on the process of designing kinetic structures rather than the product [20].

Motion Grammar is used to define the setup of robotic systems through Context-Free language. The grammar is the main composer of the motion parser in robotic motion. The parser is a program (part of the processor) that recognizes the input language of Motion Grammar, which is received as a series of instructions and performs the matching semantic rules per each production, as shown in Figure 2 [7].

Motion Grammar elements ($\mathcal{F}$, Z, U, η, V, P, K), where;

$\mathcal{F}$ sensor reading
Z tokens showing the robot status (output)
U robot commands part
η tokenizing function
V series of non-terminals
P series of productions
K set of semantic rules

The output Z is divided into tokens ζ with the main stream of motion being readable for the parser. Depending on the token sequence, the parser adopts a control action U to the robot. Token type ζ is responsible for the selection of a suitable production, while the production's semantic rules depend on continuous change in the Z value to achieve U. This means that the language of the robot's sensor reading is presented by the Motion Grammar. The conversion of the language of sensor reading into the language of controllers and controlling actuators is the Grammar's Motion Parser. Attributes are used to identify semantics of Motion Grammar. They are parameters linked to each token and non-terminal, with continuous change in relation to the continuous-valued variables as sensor values. Continuous and discrete information is transferred between the robot and controller. Discrete events are presented in η and form part of the string, while the continuous part of the controller is located in the sensor readings or semantic rules caused by attributes of tokens and non-terminals, and is saved by the parser.

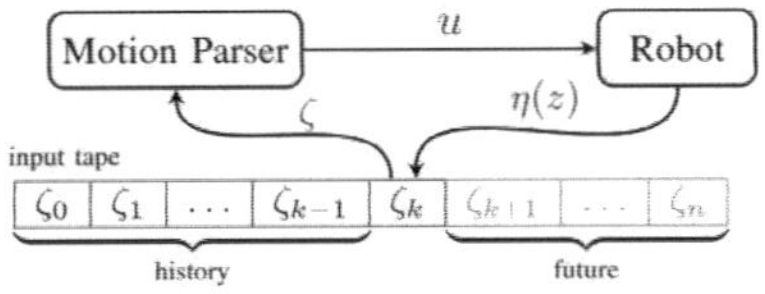

Figure 2. Process of Motion Grammar [7].

A Motion Grammar for any task depends on the nature of the task itself and the robot hardware. By applying this in the field of robotic systems, it is shown that U and Z are respectively the inputs and sensors in the robot. Token Z is designed as a set of discrete events and timeouts through the implementation of task. Tokenizing functions η need to be designed to occur from $\mathcal{F}$ to Z. The tasks are divided hierarchically into a series of easier tasks consisting of nonterminal V and productions P until a closed controlled loop is set. Through the production phase, semantic rules K for each production are obtained. Calculations for the token's attributes and nonterminal in production are done in the semantic rules K phase, till it reaches the end of the control loops, then the computed commands are transferred to the robot. Finally, to close the loop and complete the grammar, the nonterminal start variable is chosen from V as the last part in the hierarchical decomposition.

It is the purpose of this paper to demonstrate how this logic can be applied to the type of motion experienced during the process of hygroscopic behavior of wood as a natural material. Rather than defining motion as a set of controlled and well-structured transformation operations, hygroscopic behavior of wood relies on continuous variation in one important dimension, which is moisture content.

6. Analysing Programmable Material Behavior

Any basic interactive system consists of a three-component loop that consists of input, processing and output. Inputs and outputs are physical components, while processing is where action takes place. Input technology comprises sensors, cameras and tangible computing, while output technology is divided into two classes: display-based and motion-based. Processing typically reads inputs, makes decisions and actuates the output components. Interactive systems are categorized based on processing into: linear, self-regulating, or learning systems, as shown in Figure . Self-regulating and learning systems are capable of modifying their outputs according to external stimuli [1]. Based on these definitions, shape-shifting material responses and motions can fall under the category of self-regulating systems. The focus of this paper is achieving control of the output part of this loop through time to control the behaviour of shape-shifting materials.

In essence, smart materials act as if they comprise a sensing device (sensor function), and a communication network (memory function) that transmits signals from sensors to the decision making device (processing function), then finally to actuators that move the material (actuator function) [21]. The classification by Addington & Schodek demonstrates that smart materials can function as sensors and actuators in architectural buildings. For example, controlling solar radiation through the building envelope can be controlled by sensors and actuators, which can be applied by shape memory alloys, electrorestrictive and magnetorestrictive materials.

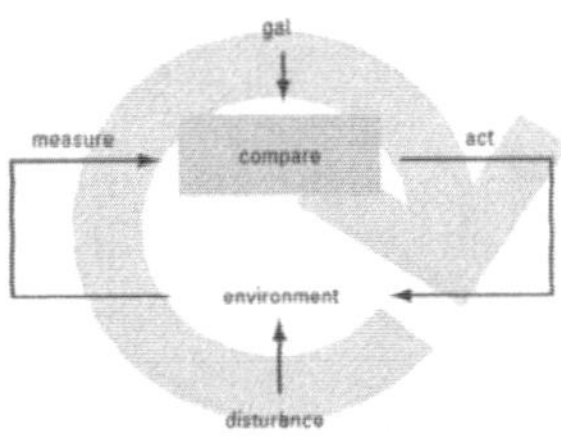

Figure 3. The basic loop of self-regulating interactive systems [1].

Kinetic motion can generally be defined through three spatial transformations: translation, rotation, and scaling, or through movement by material distortion, as in Figure. Translation is the movement of an element on a regular planar path, rotation is the element's movement around an axis, whereas scaling is the change in an object mass by expansion or contraction. Other complex movement types such as twists or rolls are a combination of these three basic movements. The fourth type of material distortion depends on material properties such as elasticity of material [22].

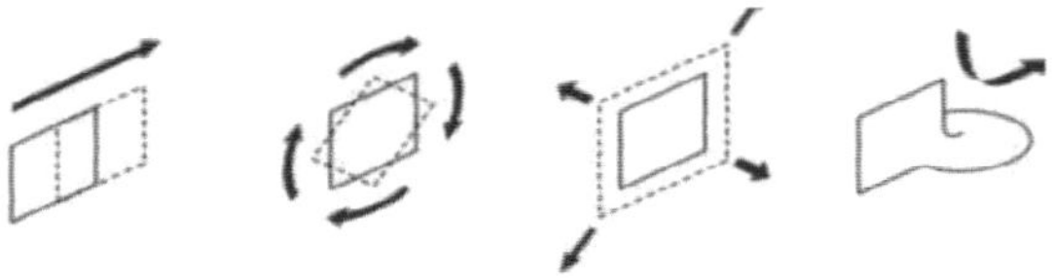

Figure 4. Kinetic motion defined through three spatial transformations (translation, rotation, and scaling) and material distortion [22].

This paper investigates the application of continuous change, using Motion Grammars, on wood as a programmable material by means of its embedded transformation properties in relation to moisture content (MC). An experiment was conducted to explore the types of exhibited motion acquired through the increase in moisture content on a rectangular sample of wood veneer. Increasing moisture content on a single side of the sample caused the wet layer to expand more than the dry layer. These unequal expansion values caused the strip to curl in relation to time. This phenomenon caused continuous and sequential differences in the deflection value (h), length (L) and radius (r) of curvature of the sample through time, as shown in Figure . A reverse motion occurs when the sample dries.

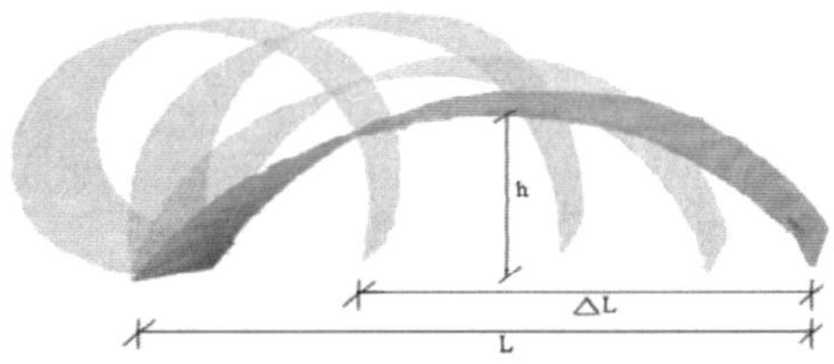

Figure 5. Sequential response motion of wood upon increase in MC.

Based on Motion Grammar logic and the motion parameters defined earlier, the following motion grammar elements can be understood accordingly as follows:
$\mathcal{F}$ refers to the values of the readings measured in this experiment through manual recordings or through sensing devices, for example the moisture content within the material sample, in addition to the current time, deflection, and length of sample measured upon applying moisture.
Z refers to the output motion, which is linked to the properties of the material, such as thickness, type, lamination, and grain orientation. This output is reflected in the values of ΔL, reflecting change in sample length; Δh, reflecting change in deflection; r, reflecting change in radius of curvature; in addition to duration of response time.
η refers to the main tokenizing function related to increase or decrease of moisture content, where moisture content defines the main driver for motion in wood.
V denotes the series of non-terminals that affect the motion outcome such as: thickness, grain orientation, type of material and lamination layers.
P refers to a series of productions that are divided into two phases. Each phase consists of discrete events located in the η tokenizing function that can be represented by a series of events. The first phase begins

due to an increase in moisture content, which results in an increase in deflection and radius of curvature, and a decrease in material length, as shown in Table. The reversible motion of the sample takes place in the second phase, which exhibits a decrease in moisture content as the sample dries, leading to a decrease in deflection and radius of curvature, and an increase in material length to return to its initial state.
K refers to the semantic rules that are directly linked in this case to the ability of the material to absorb, handle and release water, in addition to the mechanical properties of materials regarding its expansion and shrinking percentages.
U robot commands refer in this context to the overall expansion and shrinkage of wood, as a low-level action, e.g. bending the material, and flattening the material.
By defining these motion grammar elements, the framework of material motion behavior due to hygroscopic properties and change in moisture content can be identified. By programming these elements and tracking wood behavior under different conditions and using different types, thicknesses, grain orientations, and lamination mechanisms, different mechanisms can be devised to achieve desired controlled behavior.

Table 1. Defining productions, semantic rules and robot commands of wood motion in relation to motion grammar

Productions (P)	Semantic rules (K)	Robot Commands (U)
Production 1	Increase in MC	Bending the material
	Increase in Δh	
	Increase in r	
	Decrease in ΔL	
Production 2	Decrease in MC	Flattening the material
	Decrease in Δh	
	Decrease in r	
	Increase in ΔL	

7. Conclusion
This paper introduced a method to track and analyze the motion derived from programmable materials through Motion Grammar (G_M). The implication of G_M on programmable materials relies on the way the material is fabricated in which the same material acts as a sensor, processor and output element. This embedded property is enhanced through the way material is programmed. Using G_M permits an analysis method for each property and variable in this material. Tracking and analyzing this motion-embedded material allows for passive control of smart material behavior for kinetic structures in buildings.
Based on the framework in this paper, as the tokens, attributes, tokenizing functions, non-terminals, productions, semantic rules, and commands are identified, specific and customized kinematic operations can be defined to allow for material programming and to achieve a desired behavior; for example, in response to specific climatic conditions. Future development will include further investigation of motion grammar parameters for different programmable materials as a method to analyze and track their response behavior. Experimenting with wood under different conditions and climates is a direct application of the proposed framework.

8. Acknowledgment
The authors are indebted to the sponsorship provided by the Bartlett Family Fund in supporting the 'Soft Adaptive Building Skins for Energy Efficient Architecture' research project.

9. References

[1] C. Calderon, Harvard University, Graduate School of Design, *Interactive architecture design*. Cambridge, Mass.: Harvard Graduate School of Design, 2009.

[2] N. Gu and X. Wan, *Computational design methods and technologies: applications in CAD, CAM, and CAE education*. Hershey, PA: Information Science Reference, 2012.

[3] S. Nabil *et al.*, "Interioractive: Smart Materials in the Hands of Designers and Architects for Designing Interactive Interiors," in *(DIS '17)*, Edinburgh, UK, 2017.

[4] A. Nijholt and A. Minuto, "Smart material interfaces: Playful and artistic applications," in *Conference on Imaging, Vision & Pattern Recognition (icIVPR)*, Dhaka, Bangladesh, 2017, pp. 1–6.

[5] M. Fox, Ed., *Interactive architecture: adaptive world*, First edition. New York: Princeton Architectural Press, 2016.

[6] M. Kretzer, *Information materials: smart materials for adaptive architecture*. Springer International Publishing, 2017.

[7] N. Dantam and M. Stilman, "The Motion Grammar: Linguistic Perception, Planning, and Control," in *Robotics: Science and Systems VII*, Cambridge, Mass.: MIT Press, 2012.

[8] D. M. Addington and D. L. Schodek, *Smart materials and new technologies: for the architecture and design professions*. Amsterdam; Boston: Architectural Press, 2005.

[9] A. Ritter, *Smart materials in architecture, interior architecture and design*. Basel; Boston: Birkhäuser, 2007.

[10] E. Lefebvre, J. Faucheu, B. D. Curto, and D. Delafosse, "Stimuli-responsive materials: Definition, classification and descriptions - poster," in *7th International Materials Education Symposium*, Cambridge, 2015.

[11] R. M. Erb, J. S. Sander, R. Grisch, and A. R. Studart, "Self-shaping composites with programmable bioinspired microstructures," *Nat. Commun.*, vol. 4, p. 1712, Apr. 2013.

[12] D. Correa *et al.*, "3D-Printed Wood: Programming Hygroscopic Material Transformations," *3D Print. Addit. Manuf.*, vol. 2, no. 3, pp. 106–116, Sep. 2015.

[13] Forest Products Laboratory, *Wood handbook: wood as an engineering material*. Madison, Wis.: The Laboratory, 2010.

[14] A. Holstov, P. Morris, G. Farmer, and B. Bridgens, "Towards sustainable adaptive building skins with embedded hygromorphic responsiveness," presented at the Advanced Building Skins, 2015.

[15] N. Dantam, I. Essa, and M. Stilman, "Linguistic transfer of human assembly tasks to robots," 2012, pp. 237–242.

[16] Terry Knight, "Introduction to shape grammars," MIT, 07-Feb-2000.

[17] D. El-Zanfaly, "Active shapes: Introducing guidelines for designing kinetic architectural structures," in *SIGRADI 2011*, Argentina, 2011.

[18] Ardavan Bidgoli and Daniel Cardoso-Llach, "Towards a Motion Grammar for Robotic Stereotomy," in *20th International Conference of the Association for Computer-Aided Architectural Design Research in Asia (CAADRIA) 2015*, Hong Kong, 2015, pp. 723–732.

[19] W. Goddard, *Introducing the theory of computation*. Sudbury, Mass: Jones and Bartlett Publishers, 2008.

[20] D. El-Zanfaly and S. Abdelmohsen, "Imitation in Action: A Pedagogical Approach for Making Kinetic Structures," in *CAADFutures 2017*, Istanbul, 2017, pp. 533–545.

[21] S. S. lyer and Y. M. Haddad, "Intelligent materials-An overview," *Int J Pres Ves Pip.*, vol. 58, pp. 335–344, 1994.

[22] J. Moloney, *Designing kinetics for architectural facades: state change*. Abingdon, Oxon; New York: Routledge, 2011.

IOP Conf. Series: Materials Science and Engineering **362** (2018) 012012 doi:10.1088/1757-899X/362/1/012012

The Effect of Graphene Oxide on Cement Mortar

Henrik Kjaernsmo[1], Samdar Kakay[2] Kjell T Fossa[3] and John Gronli[4]

1 Student, University of Stavanger,
Norway, E-mail: henrik.mac@gmail.com
2 Senior Engineer, University of Stavanger, Norway, E-mail: Samdar.s.kakay@uis.no
3 Adjunct Professor, University of Stavanger, Norway
4 Chief Engineer, University of Stavanger, Norway

Abstract. This paper investigates the effect of water dispersed- and powder Graphene oxide (GO) nanoparticle on fresh cement mortar, microstructure and mechanical strength after 3,7, and 28 days of curing. These properties were studied by treating the cement mortar with 0.03 wt%, 0.05 wt%, and 0.2 wt% GO of the cement weight combined with 0.8wt % polycarboxylate superplasticizer. The results show that the workability decreases as increasing the content of water dispersed GO. The heat of hydration is increased for both types of GO systems. The percent air content in 0.03 wt% and 0.05 wt% GO is almost constant, but increased from 3.2 % to 4.9 % in 0.2 wt% water dispersed GO. The increased air content has effect on poor compaction and workability. GO has the potential of accelerating the hydration process and enhance the early mechanical strength (3 and 7 days), but the workability seems to diminish the mechanical strength after 28 days of curing, particularly for the highest content of water dispersed GO. No distinct influence of GO on the microstructure. The overall results showed that the impact of water dispersed GO was found out to be higher than the powder GO.

1. Introduction

In recent years, the application of nanotechnology (1-100nm) has shown an impressive results in solving several engineering related problems in electronics and biomedicine. Through chemical and physical processes, nanomaterials improve material properties such as thermal, mechanical, electrical, lubricity and rheological. Comparing with the micro/macrosized particles, the surface area to volume ratio of the nanoparticle is higher Amanullah et al, (2009) [1].

Nanomaterials improve cement properties (Ruhal et al, 2011 [2], Ershadi, et al (2011) [3], Li, et al (2003) [4], Rui et al. (2015) [5]. Among others, Graphene is known for its superb material properties and has shown a great potential in other industries, such as electronics and polymer composites. Graphene is about 200 times stronger than steel [6]. Graphene Oxide (GO) is a derived from graphene. Wang et al 2015[7] have reported the influence of GO impact of of graphene oxide on the microstructure and mechanical strength of cement, which has potential of providing a nano-reinforcement. However, the research on GO is still at an early stage. This paper presents the effect of GO in the conventional cement mortar. The investigation is through various experimental studies.

2. Experimental works

Wang et al. 2015[7], reported enhanced mechanical properties with the GO 0.03 wt.% GO and 0.05 wt% of the cement weight. In this paper, 2.00 wt.% GO concentration were also evaluated to investigate the performance of high concentration on the mortar properties. The reference (control) cement mortar was formulated without nanoparticle additive. The impact of GO nanoparticle on the reference system were studied by adding a 0.03wt.%, 0.05wt.%, 0,2wt % GO nanoparticle. This section presents the description of cement mortar additives, cement mortar formulation, measurement and characterization

2.1. Materials and test methods.

Table 1 shows cement mortar ingredients used in the reference and GO nanoparticle treated cement mortar. The mix design was according to the procedure described in *NS-EN 196-1:2016 – Methods of testing cement* [8]. The reference cement mortar was formulated by mixing 450gm standard cement, 1350gm CEN Reference Sand and 3225gm Water. The GO treated mortar was formulated by adding GO powder and GO water dispersed in the reference mortar system. The cement mortar ingredients were mixed with a Hobart N-50 mechanical mixer for 5 min.

Cement mortar formulation

1. Reference (Ref.) = 450gm cement, + 1350gm Sand + 3225gm + Water +0.8 wt.% Superplastizer
2. Graphenea (Water dispersed) = Ref + [0.03 wt.%, 0.05 wt.%, 0.2 wt.%] GO
3. Graphitene (Fine powder) = Ref + [0.03 wt.%, 0.05 wt.%, 0.2wt.%] GO

Table 1. Mix Design with different dosages of GO

Materials	Quantity [g]
Standard cement	450 ± 2
CEN Referance Sand	1350 ± 2
Water	3 225 ± 1
MasterGlenium ACE 434 Superplasticizer (SP)	0.8 wt.% of the cement weight
Graphene Oxide (GO)	[0.00, 0.03, 0.05, 0.2] wt.% of the cement weight
Graphenea & Graphitene	[0.00, 0.03, 0.05, 0.2] wt.% of the cement weight

The experimental test methods was based on *NS-EN 196-1:2016 – Methods of testing cement "Part 1: Determination of Strength" [9].* The prime objective is to investigate the effect of graphene oxide (GO) on:

- *Fresh cement mortar* (Workability, Air content, Density , and Heat of hydration)
- *Mechanical properties* (Flexural strength, Compressive strength, Splitting Tensile Strength and Ultrasonic Velocity)
- *Microstructure* (The interfacial transition zone & the bulk cement paste)

2.2. Results and discussion.

The effect of GO on the fresh mortar and on the cement specimen was investigated through the methods presented in the previous section. The results are summarized as follows.

2.2.1. Effect of GO on the fresh cement mortar

Workability

A reduced workability is one of the challenges with graphene oxide. Consequently, it can cause poor compaction and therefore influence the mechanical properties. The impaired workability can be explained by a reduced amount of free-water since GO has a significant high specific surface which can result in a high water absorbency. In literature it is documented that the reduced workability has been solved by adding superplasticizer (SP). However, the potential side-effects due to chemical reaction between SP and GO are unknown to the investigators.

The results from the mini-flow test are presented in Table 2. Results show that for the 0.03wt.%, 0.05wt% and 0.2wt.% concentration of the Graphenea (water dispersed), the mini-flow diameter of the fresh mortar decreased by 5.5%, 18.9% and 60.6%, respectively. On the other hand in the Graphitene (fine powder) system, the mini-flow diameter decreased by 3.5%, 2.8% and 3.5%, respectively. From this test results one can observe that there is a correlation between the content of Graphenea (water dispersed) and the reduced flow diameter. For Graphitene (fine powder), there is only a small change in the flow diameter with increasing content of GO, but no correlation between the percent GO and the flow diameter. The cement mortar with Graphenea 0.20% & SP 0.8% had a stiff plastic consistency after mixing. The flow diameter was measured to only 10 cm, which equal to zero flow since the bottom diameter of the cone is 10 cm, see Figure 1.

According to Gong et al. (2015) [10], the mini-flow diameter is reduced by 34.6 % with a GO content of only 0.03 wt%. Similar results are also reported by Pan et al. (2015) [11], Shang et al. (2015) [12], and Tang et al. (2014) [13]. Shang et al. (2015) [12], also observed that the mini-flow diameter decreases with an increasing content of GO. The mini-flow diameter is reduced by 36,2 % with a GO content of 0.08 wt%. According to Chuah et al. (2014) [7] [14], similar results are also observed with other nanomaterials, such as nano silica and carbon nano tubes.

Figure 1. Mini-Flow Test - 0.2% Graphenea & 2.0% SP

Air content

The air content results are presented in Table 2. Results show that for the 0.03wt.%, and 0.05wt% Graphenea increased by 0.2% concentration and in 0.2% increased by 1.7%. This can be explained by the stiff plastic consistency that was observed during the mini-flow test. On the other hand, as the Graphitene (fine powder) concentration increases (0.03wt.%, 0.05 wt.% and 0.2wt.%), the air content decreased by 0.7%, 0.7% and 0.3%, respectively

The densities of all the mortar are within the range of 2.22 to 2.3 kg/m^3. The small reduction in density for Graphenea 0.2% seems to be a result of the reduced workability, which was observed during the mini-flow test.

Heat of hydration

According to the author's knowledge, the influence of GO on the heat of hydration is not clarified well in previous research. For instance, Zhao et al. (2017) [8] [15] reported that GO increases the maximum cement hydration exothermal rate, while Wang et al. (2015) [6] observed the opposite.

Figure 2 presents the temperature developments of different mortar compositions. Graphene oxide (GO) treated cement mortar shows an earlier temperature development and a higher maximum temperature. These results indicate that graphene oxide might have chemically reacted with the cement mortar components, and hence results in increasing the exothermal heat development. The maximum (peak) temperatures with respect to the reference mortar are presented in Table 2. The results also indicate that the impact of water dispersed GO (Graphenea) on the temperature development is higher than the GO in powder form (Graphitene). For instance, in the 0.05wt.% GO additive systems, the water dispersed GO (Graphenea) and powder form (Graphitene) increased temperature development by 5.4% and 2.9%, respectively. Similarly, in the 0.2wt.%, the temperature development increased by 2.6% and 0.7%

respectively. One can observe a non-linear effects with respect to concentration. However, more studies required to come to a conclusion. There is also a correlation between the content of the water dispersed GO (Graphenea) and maximum (peak) temperature.

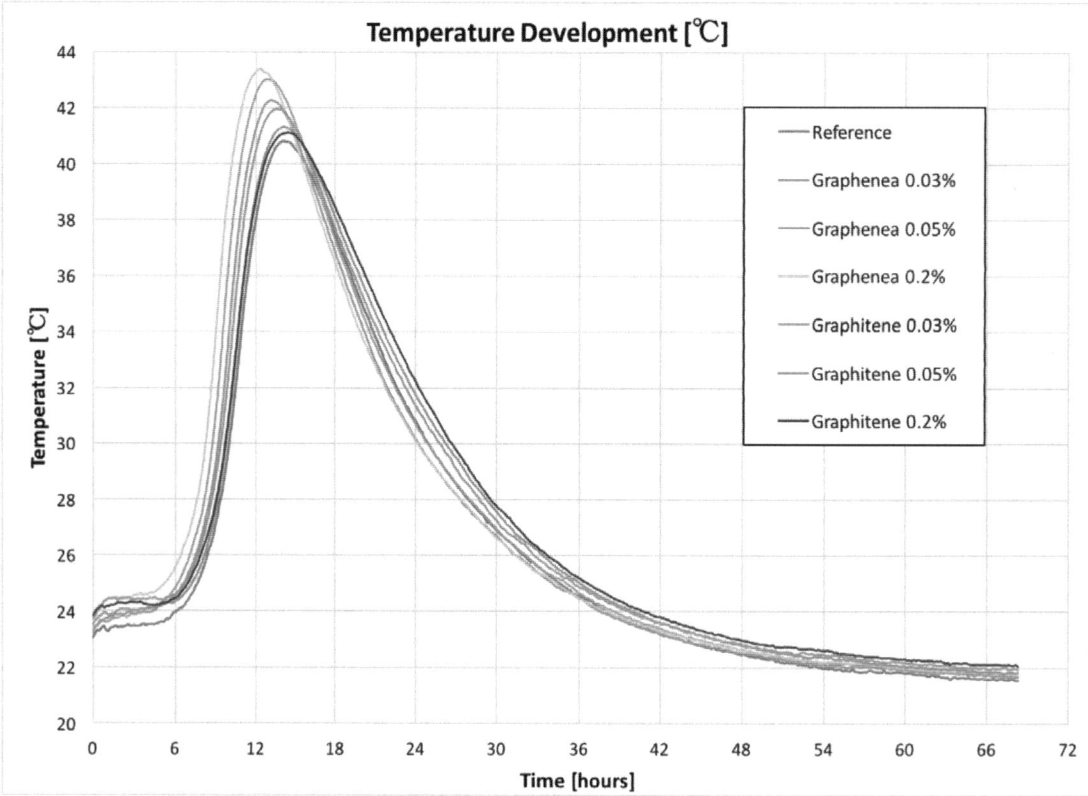

Figure 2. Temperature development during 3 days of curing

Table 2. Fresh mortar properties

Material	GO [wt.%]	SP [%]	Mini-Flow Test[cm]	Air content [%]	Density [kg/m³]	Temperature Max(peak) °C
Reference	0	0.8	25.4	3.2	2299	40.8
Graphenea	0.03	0.8	24.0	3.4	2264	42.3
Water dispersed	0.05	0.8	20.6	3.4	2262	43.0
	0.20	0.8	10.0	4.9	2223	43.4
Graphitene	0.03	0.8	24.5	2.5	2278	41.3
Fine powder	0.05	0.8	24.7	2.7	2278	42.0
	0.20	0.8	24.5	2.8	2261	41.1

2.2.2 Effect of GO on Mechanical Properties:
Flexural Strength:
The flexural strength of the GO treated and untreated mortar were measured at three, seven and 28 days of curing and the results are presented in Table 3. Graphene oxide seems to improve the early flexural strength after three and seven days, but have no effect at 28 days. No distinct variation between the content or type of graphene oxide and the flexural strength is observed. The percent flexural strength development from three to 28 days is also lower compared to the reference mortar.

The improved flexural strength after three and seven days can be a consequence of the increased heat of hydration. This indicates that graphene oxide has a potential of accelerating the hydration process which increases initial flexural strength development.

Strain Distribution – Verification of the Three-Point Bending Test

The typical strain distribution during the three-point bending test is presented in Figure 3A. The red area indicates that the strain arises and propagates in a very concentrated part of the specimen. The prism ends are not subjected to substantial strain during the three-point bending test, and therefore, validates the method of using the prism ends to determine the compressive strength.

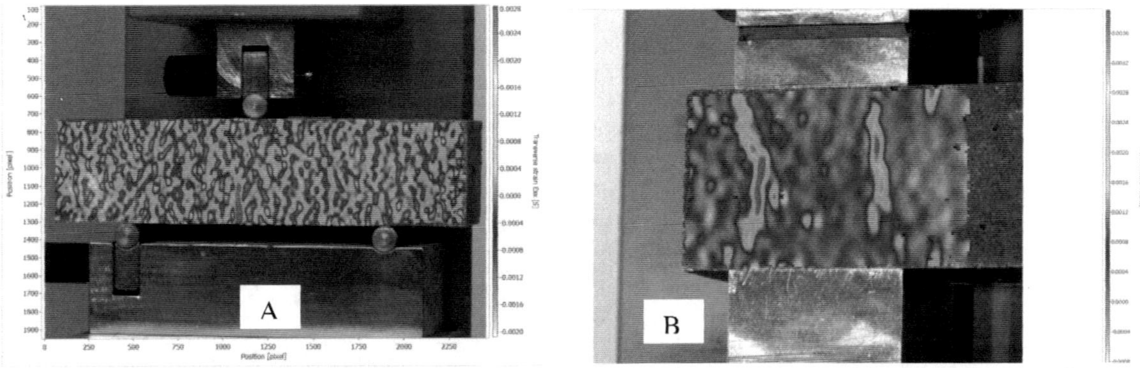

Figure 3: Three-Point Bending (A) and Compressive (B) Test Transverse Strain distribution in x-direction

Compressive strength

The compressive strength after three, seven and 28 days of curing is presented in Table 3. Graphene oxide increases the early compressive strength after three and seven days, but have no effect after 28 days of curing. In addition, the compressive strength for Graphenae 0.2% is reduced by 8 % after 28 days, while compressive strength for Graphenea 0.2% & SP 2.0% is improved by 7 % after 28 days. This indicates that the reduced workability has a considerable effect on the compressive strength and can result in a variation of 15 %.

Strain Distribution – Verification of the Compressive Strength Test

The strain distribution during the compressive strength test is presented in Table 3. The red zones, which indicate a relatively higher substantial strain compared to other parts of the specimen, are first initiated in the middle of the specimen, as illustrated in Figure 3B. The strain distribution propagates then towards the bottom and the top part. The observed strain distribution and the propagation is equivalent to a traditional compressive strength test with cubes, and the selected method of determining the compressive strength can be considered as valid.

Splitting Tensile Strength Results

The splitting tensile strength results compared to reference mortar are presented in Table 3. The results show an enhanced splitting tensile strength after three and seven days of curing for the cement mortar containing Graphenea. The highest enhancement is 19 % for the cement mortar containing 0.2 wt.% Graphenea and after three days of curing. Moreover, this enhancement seems to decline during the hydration process, and at 28 days the splitting tensile strengths are almost equal or less than the reference mortar. The results are almost equivalent to the reference mortar for the cement mortar containing 0.03wt.% Graphitene. For the other mortar compositions containing Graphitene, the splitting tensile strength is reduced by 6 % and 8 % at 28 days. The results indicate that graphene oxide produced by Graphena can accelerate the strength development in the early stages of the hydration process. The strength development seems to decline after three days of curing and no improvements are obtained at 28 days.

Strain Distribution – Verification of the Splitting Tensile Test

According to the theoretical cylinder stress distribution, the top and bottom part of the cylinder face are subjected to compression, while the middle is exposed to tension. Figure 4 presents the horizontal strain evolution on the cylinder face during the splitting tensile test. The image shows no substantial strain near the edges of the cylinder, while the red areas in the middle indicate strain. The strain evolution arises first in the middle of cylinder face and propagates outwards to the cylinder edge. These observations can be associated with the theoretical cylinder stress distribution. However, it is important to emphasize that the presented theory of the splitting tensile test is related to stress through the whole cylinder, and the images express the actual strain distribution observed on the cylinder face.

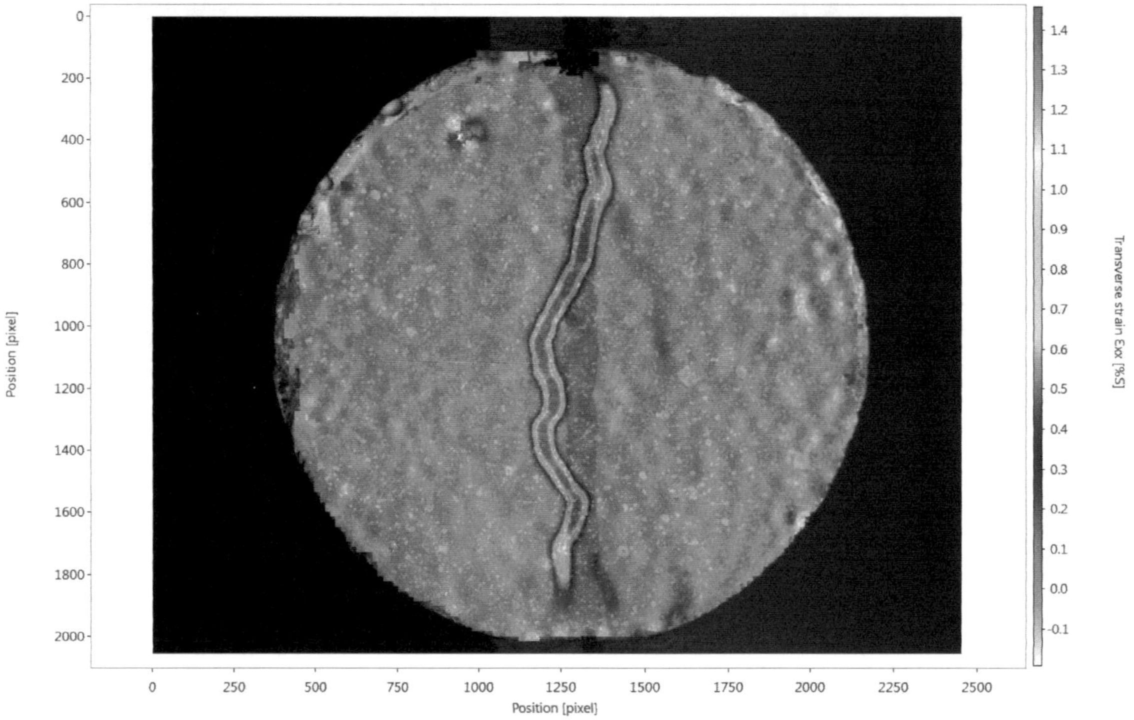

Figure 4: Strain evolution (ε_{xx}) during the splitting tensile test

Three virtual strain gauges were arranged as illustrated in Figure 5. The corresponding strain evolutions are presented in Figure 6. The virtual strain gauge positioned in the center of cylinder face (red) show substantial larger strain compared to those located close to the edges (blue & green). Moreover, the bottom strain gauge (green) shows a negative strain in the earlier stages of the strain evolution. These results indicate that the selected setup for the splitting tensile strength test provides equivalent stresses as for a conventional splitting tensile test with a larger cylinder and different loading rate.

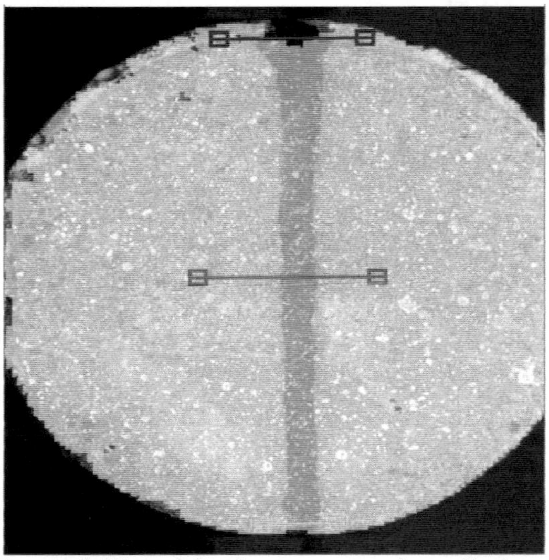

Figure 5. Virtual Strain gauges on the cylinder

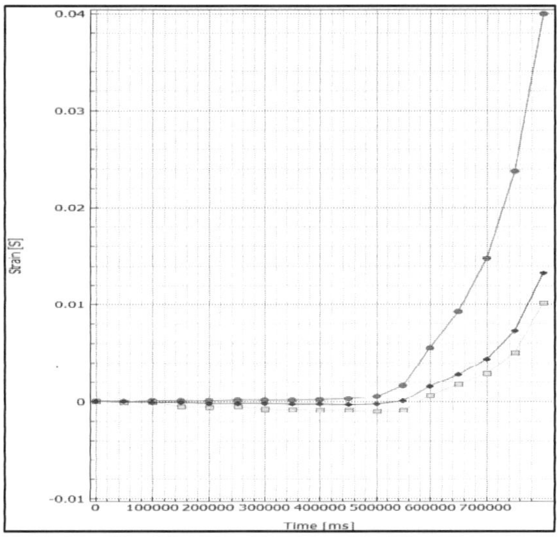

Figure 6. Strain evolution corresponding to the face strain gauge

Ultrasonic Velocity Test

The ultrasonic velocity test is non-destructive method, which determines a special type of modulus of elasticity, and should not be confused with the more conventional E-modulus or Young´s Modulus (i.e the ratio of stress to strain). The sonic velocity is determined from the travel time and the length of cement plug. The modulus of elasticity is calculated as [16]:

$$M = \left(K + \frac{4}{3}.G\right) = V_p^2.\rho \tag{1}$$

$$Volume = \pi.r^2.L \tag{2}$$

$$\rho = \frac{mass}{volume} \tag{3}$$

$$V_p = \frac{L}{\Delta t} = \frac{10^{-6}.\frac{L}{1000}}{\Delta t} \tag{4}$$

Where M is the modulus of elasticity, K is the bulk modulus and G the is shear modulus, ρ is density and V_p is compressional sonic velocity.

The Modulus of Elasticity (M)

The modulus of elasticity (M) after seven and 28 days of curing is presented in Table 3. The results show a reduced modulus of elasticity for the mortar cylinders containing Graphenea, especially at 28 days of curing.

Detecting embedded air voids through sonic velocity

The ultrasonic velocity test results can also indicate how the reduced workability influences the amount of embedded air voids in the mortar. Since the sonic sound is not transmitted through air bubbles and voids, the sonic travel time will increase if large air bubbles or voids are located in the sonic travel path [4].Reduced sonic velocity indicates possible embedded air voids. The sonic test results take the density into account, and can therefore not be used to quantify possible air voids.

Sonic Velocity

The sonic velocity is reduced for the mortar cylinders containing Graphenea, which indicates a higher content of embedded air bubbles and voids. Furthermore, the results from the density of hardened mortar and the mini-flow test indicates both directly and indirectly similar observations.

Table 3. Mechanical hardened mortar properties

Material	I %	Flexural Strength [MPa]			Compressive Strength [MPa]			Splitting T. strength [MPa]			Hardened Density [kg/m^3]	E modulus [GPa]	
Days		**3**	**7**	**28**	**3**	**7**	**28**	**3**	**7**	**28**		**7**	**28**
Reference	0	6.1	6.9	7.8	42	48	61	3.2	3.9	4.9	2364	44.2	46.5
Graphenea Water disp.	0.03	7.2	7.4	6.8	45	52	58	3.4	3.9	4.7	2348	42.7	43.6
	0.05	6.6	7.3	7.9	45	51	59	3.4	3.9	4.5	2340	43.7	41.6
	0.20	6.9	7.5	7.7	43	48	56	3.8	4.1	5.0	2282	41.6	42.5
Graophitene Fine powder	0.03	6.6	7.2	7.6	42	51	60	3.1	3.8	4.9	2362	44.9	47.6
	0.05	7.1	7.3	7.9	43	51	61	3.0	3.6	4.7	2361	45.2	48.4
	0.20	6.6	6.8	7.5	43	50	57	3.0	3.7	4.5	2362	41.6	48.6

2.2.3 Effect of GO on the Microstructure – SEM & EDS

Graphene oxide sheets were only observed by the scanning electron microscope (SEM) in a sample taken from the cement mortar containing 0.05wt.% Graphenea. The low dosages of GO (0.03 wt.%, 0.05 wt.%, and 0.2 wt.% of cement weight) and the small sample size can explain why graphene oxide sheets were not observed in the other mortar compositions. The graphene oxide sheets were verified by analyzing the bright objects with the Energy Dispersive X-ray spectroscopy. The analyzed bright objects were selected because they did not look like traditional hydration products and because of their particular shape.

The graphene oxide sheets shown Figure 7 are observed at different locations on the sample. The size of the graphene (width or length) varies approximately from 0,865 μm to 16,7 μm (diameter of a human hair is approximately 17 to 181 μm). Figure 7 shows a wrinkled graphene oxide sheet located next to a small air bubble. The percent element composition is presented in Table 4. The element analysis shows a carbon content of 64.61 % and 16.34 % oxygen. The element composition indicates with a high probability that the observed object is a graphene oxide sheet or sheets. Figure 8 and Figure 9 illustrate similar observations as in Figure 7 with element analyses (Table 5 & Table 6). Moreover, the GO sheet observed in Figure 8 has a more distinct polyhedron shape compared to Figure 7. The smallest observed graphene oxide sheet is illustrated in Figure 9. Moreover, similar wrinkled-shaped graphene oxide sheets have also been observed in previous research, but the element analysis showed a larger content of calcium (Ca).

No distinct variation in the interfacial transition zone (ITZ) between the different samples were observed, except for the graphene oxide sheet located in the ITZ for the cement mortar containing 0.05wt.% Graphenea. This is illustrated in Figure 8.

IOP Conf. Series: Materials Science and Engineering **362** (2018) 012012 doi:10.1088/1757-899X/362/1/012012

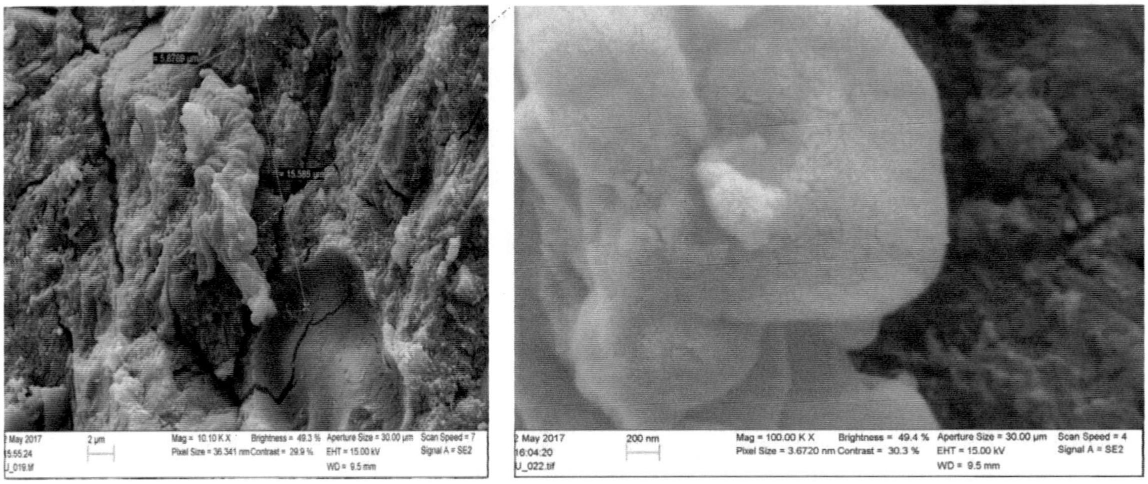

Figure 7. SEM picture of cement specimen I

Table 4. Element Analysis

Element	wt.%	At %
C K	64.61	76.30
O K	16.36	14.50
NaK	6.30	3.88
MgK	0.74	0.43
AlK	0.73	0.38
SiK	3.00	1.52
S K	1.34	0.59
CaK	6.27	2.22
FeK	0.65	0.16
Total	100.00	100.00

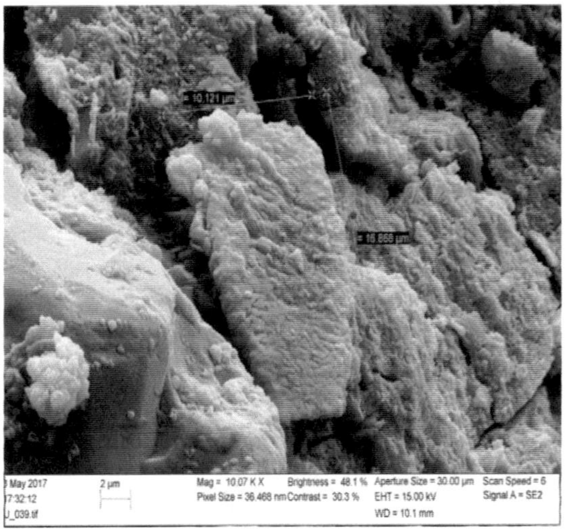

Figure 8. Graphene Oxide Sheet located in the ITZ & Spectrum II

Figure 9. The smallest observed Graphene Oxide sheet. Spectrum III is located in the center of the bright object

Table 5. Spectrum II - Element Analysis

Element	wt. %	At %
C K	73.89	83.03
O K	11.89	10.03
NaK	5.95	3.49
MgK	0.98	0.54
AlK	0.97	0.48
SiK	0.88	0.42
S K	2.73	1.15
CaK	1.97	0.66
FeK	0.74	0.18
Total	100.00	100.00

Table 6. Spectrum III - Element Analysis

Element	wt. %	At %
C K	55.74	70.05
O K	18.57	17.52
NaK	5.40	3.54
MgK	0.90	0.56
AlK	1.19	0.67
SiK	4.97	2.67
S K	1.60	0.75
CaK	10.31	3.88
FeK	1.32	0.36
Total	100.00	100.00

3. Summary

In this paper, the effects of graphene oxide (GO) on the cement mortar were studied through experimental work. The overall results showed that the impact of water dispersed GO was found out to be higher than the powder GO. The main results are summarized as follows.

- The workability is reduced by the increasing content of water dispersed GO, but is less affected by fine powder GO.
- The consistency of fresh mortar with the highest content of water dispersed GO (0.2 wt.%) is categorized as stiff plastic.
- GO has almost no effect on the content of air in the fresh mortar, except for the cement mortar with the highest content of water dispersed GO (0.2 wt.%)
- The 0.2 wt.% water dispersed GO and powder GO decreased the density of the mortar by 3,5% and 1,7% respectively
- The 0.2 wt% water dispersed GO increased the heat of hydration than the other concentration.
- GO enhances the flexural strength after three and seven days, but has no effect at 28 days. The results show no correlation between the content of GO and the flexural strength. Moreover, the highest improved flexural strengths were in the range of 9-17% and observed after three days of curing.
- The effect of GO on the compressive strength corresponds to the flexural strength. Except for 7 % improvement for the cement mortar with 0.2 wt.% water dispersed GO and a content of polycarboxylate which provides an equivalent workability as the reference mortar (additional mortar composition).
- Water dispersed GO increases the splitting tensile strength after three days of curing, but the enhancement decreases after seven days, and no effect at 28 days of curing. In addition, the powder GO has almost no influence on the splitting tensile strength at three, seven and 28 days.
- No distinct variations in the dynamic modulus of elasticity are detected after seven days of curing. At 28 days of curing the water dispersed GO reduces the dynamic modulus of elasticity, and the opposite effect is observed for fine powder GO. Moreover, the reduced sonic velocity for cement mortar with water dispersed GO indicates an increased content of air voids.
- SEM and X-ray detector have shown that Graphene oxide sheets size varies from 0,865 μm to 16,7 μm. The element analysis showed a high content of carbon (55.74 - 73.89 wt.%) and oxygen (11.89 – 18.57 wt.%). Furthermore, no distinct change in the interfacial transition zone or the bulk cement paste have been observed, except for the presence of GO sheets.

4. References

[1] Md. Amanullah, and Ashraf M. Al-Tahini: (2009)// *Nano-Technology-Its Significance in Smart Fluid Development for Oil and Gas Field Application*// SPE 126102 Alkhobar, Saudi Arabia, 09-11 May 2009

[2] Rahul C. Patil, Abhimanyu Deshpande 2012 *//Use of Nanomaterials in Cementing Applications//* SPE-155607, 12-14 June, Noordwijk, The Netherlands 2012
[3] Ershadi, V. et al (2011) *The Effect of Nano silica on Cement Matrix Permeability in Oil Well to Decrease the Pollution of Receptive Environment. International Journal of Environmental Science and Development,* Vol. 2, No. 2, April 2011.
[4] Li, H. et al (2003) *Microstructure of cement mortar with nano-particles. Composites: Part B* 35 (2004) 185–189.
[5] Rui Zhang, Xin Cheng, Pengkun Hou, Zhengmao Ye, *//Influences of nano-TiO2 on the properties of cement-based materials: Hydration and drying shrinkage, Construction and Building Materials*, Volume 81, 15 April 2015, Pages 35-41, ISSN 0950-0618
[6] Vjayaraghavan, D.A. *How strong is graphene?* 2017 [cited 2017 May 21, 2017]; Available from: http://www.graphene.manchester.ac.uk/discover/video-gallery/what-is-graphene/how-strong-is-graphene/.
[7] Wang, Q., et al., *Influence of graphene oxide additions on the microstructure and mechanical strength of cement.* New Carbon Materials, 2015. **30**(4): p. 349-356.
[8] NS-EN 196-1:2016 – Methods of testing cement.
[9] NS-EN 196-1:2016 – Methods of testing cement "Part 1: Determination of Strength"
[10] Gong, K., et al., *Reinforcing Effects of Graphene Oxide on Portland Cement Paste.* Journal of Materials in Civil Engineering, 2015. **27**(2).
[11] Pan, Z., et al., *Mechanical properties and microstructure of a graphene oxide-cement composite.* Cement and Concrete Composites, 2015.
[12] Shang, Y., et al., *Effect of graphene oxide on the rheological properties of cement pastes.* Construction and Building Materials, 2015. **96**: p. 20-28.
[13] Tang, L., et al., *Pre-Study of Graphene-Enhanced Cementitious Materials.* 2014, Chalmers University of Technology: Chalmers University of Technology, Gothenburg.
[14] Chuah, S., et al., *Nano reinforced cement and concrete composites and new perspective from graphene oxide.* Construction and Building Materials, 2014. **73**: p. 113-124.
[15] Zhao, L., et al., *Mechanical behavior and toughening mechanism of polycarboxylate superplasticizer modified graphene oxide reinforced cement composites.* Composites Part B: Engineering, 2017. **113**: p. 308-316.
[16] Sheriff, R.E. and L.P. Geldart, *Exploration Seismology.* Second ed. 1995, The Pitt Building, Trumpington Street, Cambridge CB2 1RP, United Kingdom Cambridge University Press.

Acknowledgments
This research was financed by University of Stavanger (UiS), Norway, through a master thesis *(Title: Graphene Oxide – A New Potential Nano Reinforcement*) written by Henrik Eeg Kjaernsmo (1st author of this paper), Spring 2017. Additionally, we would like to offer our special gratitude to Associate Professor Mesfin BelaynehAgonafir (UiS, Petroleum Department) and Associate Professor Kidane Fanta Gebremariam (UiS, Museum of Archaeology), for their support and engagement in this research. A great appreciation also goes to laboratory staff members in the Concrete- and SEM laboratory (UiS).

Study on the Connecting Length of CFRP

Xiongfei Liu[1,a*], Yue Li[2,b], Zhanguo Li[3,c]

1 The Key Laboratory of Urban Security and Disaster Engineering, MOE, Beijing Key Lab of Earthquake Engineering and Structural Retrofit, Beijing University of Technology, Beijing 100124, China
E-mail: a:liuxfking@foxmail.com, b: liyue@bjut.edu.cn, c:lizg@bjut.edu.cn

Abstract. The paper studied the varying mode of shear stress in the connecting zone of CFRP. Using epoxy resin (EP) as bond material, performance of specimens with different connecting length of CFRP was tested to obtain the conclusion. CFRP-confined concrete column was tested subsequently to verify the conclusion. The results show that: (1) The binding properties of modified epoxy resin with CFRP is good; (2) As the connecting length increased, the ultimate tensile strength of CFRP increased as well in the range of the experiment parameters; (3) Tensile strength of CFRP can reach the ultimate strength when the connecting length is 90mm;(4) The connecting length of 90mm of CFRP meet the reinforcement requirements.

1. Introduction:

Reduction in the useful service-life of the modern reinforced concrete construction, a major problem confronting the construction industry worldwide, was affected by many factors. Rather than damaging the buildings and rebuilding, using corresponding repair materials which possesses superior adhesion、fracture toughness and compatibility with a concrete substrate to repair and reinforce the reinforcement concrete structure is a more cost-effective and easily achieved method to improve the service life of the building[1]. Choosing the proper repair materials is one of the key technology to ensure the quality of the repair of cement concrete construction[2]. Epoxy resin has become one of the most commonly used repair materials for repairing construction because of its excellent mechanical properties, interfacial adhesion and durability [3,4].

There are many kinds of repair materials commonly used for mending the building construction, such as cement, epoxy resin, polyester resin, polymer latex and polyvinyl acetate, of which epoxy resin is most widely used[5-7], and it has attracted considerable research attention. To improve the performance of epoxy resin [8], polymer such as epoxy emulsion, dispersible adhesive powder and water soluble copolymer have been developed by using emulsification and particle redispersion technology.

In order to enrich the research of fiber composite material reinforced concrete columns and save the reinforcement material under the same strengthening effect, in this paper, experiments were conducted to research the varying mode of shear stress in the location of CFRP connecting zone. To verify the conclusion obtained by above experiment, a reinforced concrete column (size Φ100×200(mm)) with CRFP was tested.

2. Experiment

2.1 Raw materials

CFRP and C30 concrete were used in this research. The adhesive was modified epoxy resin(EP), which included epoxy resin, curing agent, diluent, filler, coupling agent, anti-forming agent. The main properties of the CFRP, EP and concrete are shown in Table 1.

Table 1. The main properties of the CFRP, modified epoxy resin and concrete

Materials	f/MPa	f_{max}/MPa	Δ/%	E/GPa	t/mm
Concrete	19.36	—	—	28.9	—
EP	—	48.03	1.74	98.1	—
CFRP	—	1290.32	1.80	210	0.19

Note: f is compressive strength of core of concrete column Φ100×200 (mm); $f_{max}$is the ultimate strength; Δ is the elongation; E is the elasticity modulus; t is the thickness of the CFRP.

2.2 Design of experiment

The tensile strength of CFRP with 30mm wide and 0/15/30/60/90mm long were tested according to "Technical specimen for strengthening concrete structures with carbon fiber reinforced polymer laminate" (CECS146 : 2003). To avoid the end cracking, aluminum sheet was bonded at two ends of the CFRP in the range of 35mm, as shown in Figure1. The concrete column was reinforced and compressed based on the results of the CFRP connecting test. The sketches of the reinforced columns which were sized Φ100 by 200 millimeters were shown in Figure 2.

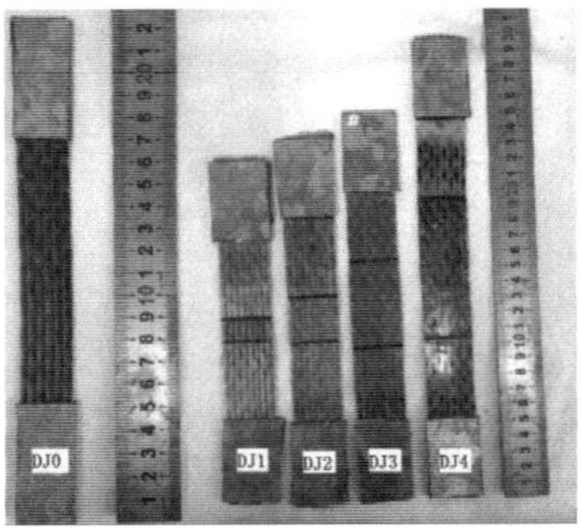

Figure 1. The specimens of CFRP

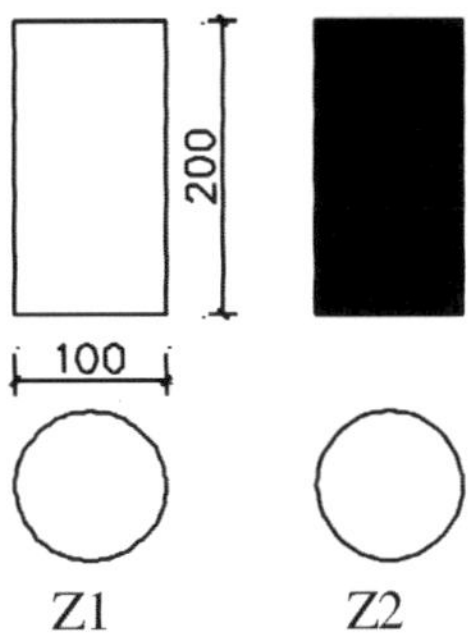

Figure 2. The dimension of the reinforced columns

3. Results and discussions

3.1 The test of CFRP

The final failure mode of specimens was shown in Figure 3.

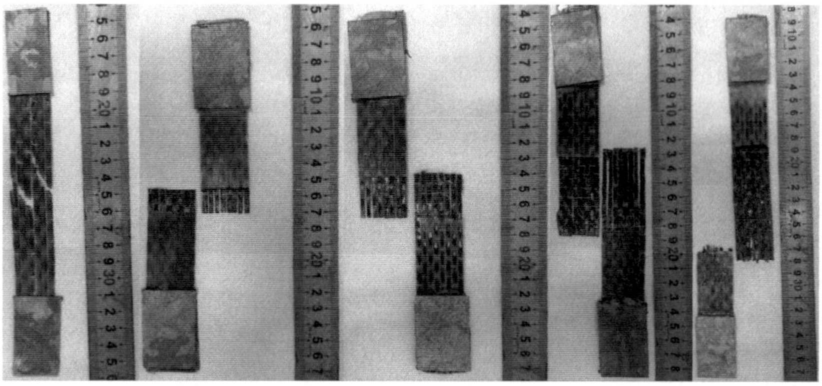

a)DJ0 b)DJ1 c)DJ2 d)DJ3 e)DJ4

Figure 3. Failure models of CFRP

From the observation of Figure 3, the appearance of CFRP had no obvious changes at the preliminary stage of loading. As the load increased, cracks occurred in the EP connecting zone. When subjected to ultimate load, specimens of DJ 1~3 damaged due to sheer stress at the bottom of CFRP. DJ4 and DJ0 crackled and failed suddenly. The test results of the connecting CFRP were shown in Table 2. Where, CFRP-The failure of the CFRP; Middle-The failure in the lapping zone of CFRP.

Table 2. Summary of the experimental results of the connecting CFRP

Code	f_u/MPa	Failure
DJ0	1290.32	CFRP
DJ1	622.58	Middle
DJ2	896.77	Middle
DJ3	1143.55	Middle
DJ4	1295.16	CFRP

Based on Table 2, the following conclusion can be obtained:
There is a linear growth relationship between the ultimate load value and the length of connecting zone of CFRP. When the connecting length is shorter than 90mm, the failure of CFRP always starts from the sheer force in the connecting zone, which means the CFRP can't reach its tensile strength before failure. When the connecting length is longer than 90mm, the failure of CFRP starts from the tension fracture of its middle region, which means CFRP reaches its tensile strength before failure, and therefore can be fully used.

3.2 The test of the reinforced columns

The final failure of specimen was shown in Figure 4.

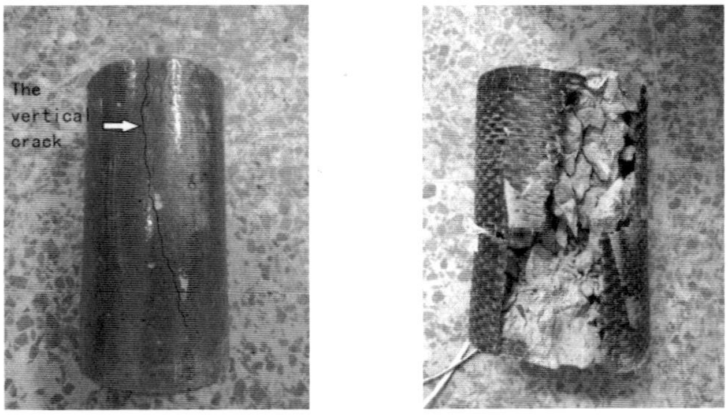

Figure 4. Failure models of columns

From the observation of Figure 4, the appearance of unconfined columns had no obvious changes at the preliminary stage of loading. As the load increased, the vertical micro-cracks developed accordingly. Under the ultimate bearing capacity of the columns, the width of the cracks got larger and the bearing capacity decreased rapidly. The failure of unconfined column was sudden and showed an obvious brittleness destruction. The appearance of entirely wrapped columns also had no obvious changes at the preliminary stage of loading. As the load increased, the axial compression deformation of the column was obvious and many micro-cracks occurred in the concrete. The sections of columns were restrained by CFRP uniformly, and all the cross sections of concrete were under the three-dimensional compressive stress. The columns stored a large deformation energy under the ultimate bearing capacity of the entire wrapping columns, and the carbon fiber at the middle position broke abruptly with a huge detonation

The compressive strength of reinforced column increased by 107.2% with a thick layer of concrete in the fracture section of CFRP, which indicated that the bond properties of CFRP and the concrete were good. The failure mode of the confined column is tear failure of the CFRP, and the connecting zone kept intact, which indicated that CFRP can reach the ultimate tensile strength meet the requirement to reinforce structure when the connecting length is over 90mm.

3.3 Micro-Analysis

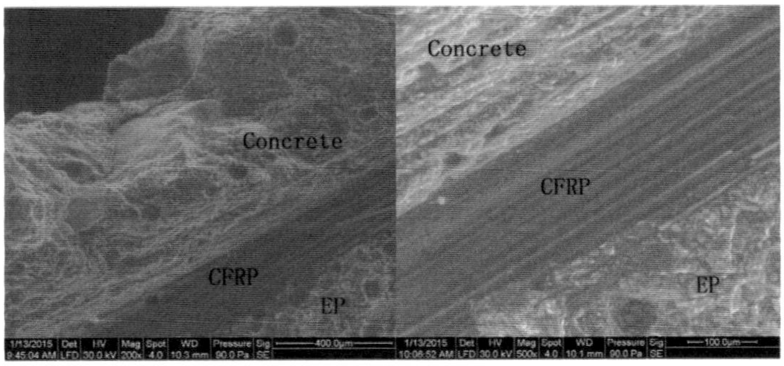

Figure 5. The fracture section images of sample

The fracture morphology analysis in Figure 5 shows that the binding properties of modified epoxy resin with CFRP was good and the interface of CFRP、concrete and CFRP was dense. Penetrating into the concrete and CFRP, EP jointed CFRP and concrete to form a structure and improved the bearing capacity of column greatly, so there is a thick layer of concrete in the fracture section of CFRP.

4. Conclusions

1)The binding properties of modified epoxy resin with CFRP is good. With EP penetrating into the concrete and CFRP, EP joints CFRP and concrete to form a structure and improve the bearing capacity of column greatly.

2)As the connecting length increased, the ultimate tensile strength of CFRP increased as well, and the tensile strength of CFRP can reach the ultimate strength with the connecting length of CFRP over 90mm.

3)The compressive strength of reinforced concrete column increased by 107.2% with 90mm connecting length, and failure didn't result from the connecting zone, which indicated the connecting length of 90mm met the requirements to reinforce structure.

5. Reference

[1] Alzahrani M M, Maslehuddin M, Aldulaijan S U, Ibrahim M. 2003 *Mechanical properties and durability characteristics of polymer- and cement-based repair materials.* Cem Concr Compos pp 527–37.

[2] Zhang Weiping. 2008 *Experimental study on chloride permeability of concrete surface treated with epoxy resin.* Journal of Building Material pp 339-344 .

[3] S. Zhong, H. Yuan. 2003 *Application of polymer in concrete,* Chemistry and Industry Publishing, Beijing. in Chinese.

[4] Ohama Y. 1997 *Recent progress in concrete-polymer composites.* Advanced Cement Based Materials pp 31-40.

[5] Dehwah HAF. 1990 *Durability of reinforced concrete beams repaired with various repair materials.* M.S. Thesis. Dhahran, Saudi Arabia: King Fahd University of Petroleum and Minerals.

[6] Shaw JDN. 1985 *Materials for concrete repair.* In: Proceedings of 1st International Conference on Deterioration and Repair of Rein-forced Concrete in the Arabian Gulf, Bahrain pp 127–38.

[7] Allen R T L. 1993 *The repair of concrete structures.* London: Blackie academic & professional.

[8] Luo J, Li Q, Zhao T. 2013 *Bonding and toughness properties of PVA fibre reinforced aqueous epoxy resin cement repair mortar.* Construction and Building Materials pp 766-771.

Acknowledgments

The authors would like to acknowledge the financial support provided by National Natural Science Foundation of China (51678011); The Importation and Development of High-Caliber Talents Project of Beijing Municipal Institutions (CIT&TCD20150310).

IOP Conf. Series: Materials Science and Engineering **362** (2018) 012014 doi:10.1088/1757-899X/362/1/012014

The Impact of Temperatures on the Stability of Rocks Surrounding a Single Fracture

Yan ZHANG [1*], Ning LI [2] and Jun DAI [1]

1 Institute of Geotechnical Engineering, Xi'an University of Science and Technology, Xi'an, Shaanxi 710054, China
2 School of civil engineering and architecture, Xian University of Technology, Xi'an, Shaanxi 710048, China
*E-mail: ylozy@126.com

Abstract. Research on the influence of temperature and the accompanying stress on the stability of the rocks surrounding an underground tunnel has become ever more important. This paper constructs a geometric model of a single-fracture tunnel by combining a high-temperature underground tunnel as the object of study with an example that uses a high-temperature tunnel segment in the water diversion tunnel of a hydropower station in Xinjiang. Based on the relevant theoretical analysis, with the consideration of different working conditions, a numerical experimental analysis was conducted to determine the two-dimensional transient temperature field distribution of the tunnel rock mass by using a numerical analysis software. The experimental data was consistent with the measured data. The calculated results show the following: a. when the temperature difference is greater, the stress concentration is higher near the fracture of the surrounding rock; b. the degree of the stress concentration in the crack tip region is not positively correlated to the distance, and there is a sensitive region where the stress varies.

1. Introduction

At present, regarding the stability of underground tunnel engineering, the effects of temperature are largely ignored. When the rocks surrounding an underground tunnel go through deformation under the action of temperature, the presence of many particles, pores and micro-cracks of low strength within the rocks will result in a large local stress concentration, which thereby induces the formation, development, compaction and interconnection of larger cracks until the entire tunnel becomes unstable and damaged.

The research on high-temperature tunnels is still not fully developed, both domestically and abroad. Reports of a high-ground-temperature phenomenon in the literature can be traced back to the second half of the 19th century. Issues related to high temperatures have occurred diversely in the construction of deep, long tunnels in various countries thereafter. In response to the frequently occurring mechanical behavior of rock under the action of high temperatures, scholars from all over the world have carried out research [1-3]. Dwivedi et al. [4] studied the physical and mechanical properties, permeability and fracture evolution pattern of Indian granite within the temperature range of 30~160°C. The basic equation of solid hydrothermal coupling for saturated fractured rocks was presented for the first time by Noorishad et al. [5]. Li Ning et al. [6] have initiated many valuable research studies; they derived three-field coupling differential equations for saturated and quasi-saturated porous fractured rocks and soil based on Gatmiri's three-field coupling study and constructed the coupling numerical model of rock, fracture, water and gas. The studies above all focus on the thermal characteristics of rock or the multi-field coupling constitutive relationship of fractured rock and soil to evaluate the rock stability, which rarely involved the stress and deformation of fracture-containing surrounding rocks under different temperatures; however, this work is particularly important and indispensable for guiding on-site construction. This paper developed a computational model for fracture-containing surrounding rocks at different temperatures by applying a combination of specific engineering with the finite element method of the transient temperature field and analyzed the distribution pattern of the stress field and deformation field for the rock mass surrounding a tunnel, which provides future design and construction with an effective theoretical foundation.

2. Computational model and boundary conditions

The water diversion tunnel of a hydropower station in Xinjiang was selected as the research subject for this paper. The surrounding rocks of the power tunnel construction adits are mainly granite-based. The temperature of the surrounding rock is with an average greater than 30°C and the highest exceeding 80°C. The depth of the granite main section in the tunnel is relatively large; thus, a high-temperature phenomenon was evidently present.

The computational model is constructed by selecting a tunnel surface containing a crack as the object of study. The numerical model of the tunnel and the application of the boundary conditions are shown in Figure 1 and Figure 2, and the diameter of the tunnel is D=6 m. According to the symmetry of the structure and load, only one-quarter of the crack length is required in the construction of the model. The length of the crack is a=0.5 m. It is assumed that the crack is a straight strip-shaped slit, the crack width is constant, the crack surface is smooth, and the crack length is far greater than the crack width. Due to the stress concentration that can occur on the crack tip in the meshing division. The network model consists of 1604 units and 1638 nodes. We applied uniform temperature to the entire model to increase the reference temperature and then used a stepped load while applying the constant temperature boundary. Kuna M, Liesk H. [7] noted that the maximum tensile stress occurred on the specimen surface where the thermal shock was received. In addition, the thermal shock is an unstable heat conduction process, the internal temperature of the specimen varies with time, and the stress induced by the thermal shock also varies accordingly. In the process of solving, the Jacobian conjugate gradient iterative equation solver is set as the solution option.

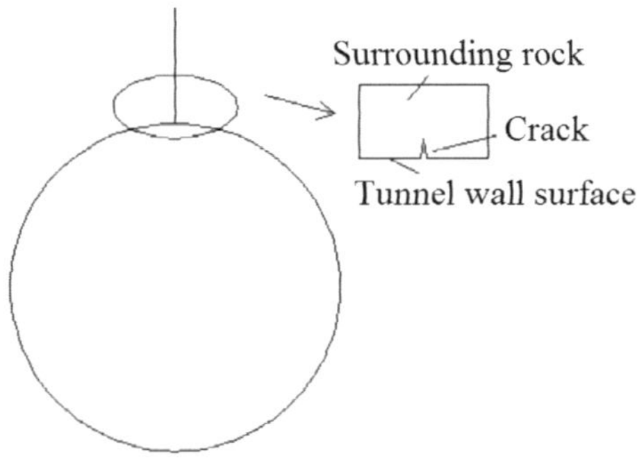

Figure 1. Vertical distribution of cracks on the roof arch and generalized model

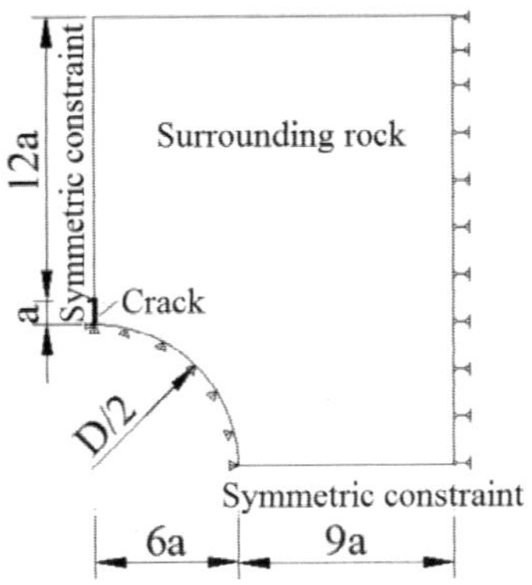

Figure 2. Diagram for the 1/4 Computational model of a crack-containing tunnel and boundary conditions

3. Determination of test plans and parameter selection

To study the influence of different temperatures on the stability of single-fracture surrounding rocks, this paper examined the impact pattern of the temperature in different regions near the fracture by taking various temperatures of surrounding rocks into consideration, starting with the fracture temperature of 5°C. The model selects the stress characteristics of the points in different regions to represent the stress received in that region, with the goal of obtaining the transferring pattern between the fracture and the surrounding rock. The numerical experiment selects granite (Ⅳ1 type) as the object. The thermal physical parameters of the material are shown in Table 1.

Table 1. Numerical experiment parameters.

Crack temperature (℃)	Tempe -rature (℃)	Deformation modulus (GPa)	Poisson's ratio	Linear expansion coefficient (10^{-6}℃)	Specific heat (J/kg ℃)	Thermal conductivity Coefficient W/(m· ℃)	Bulk weight (kN/m^3)	Depth (m)	Tunnel radius (m)
5	25	4.0	0.21	4.9	760	2.5	27	600	6.0
	60	4.2		5.3		2.4			
	70	4.5		6.2		2.3			
	80	4.7		7.5		2.2			
	90	4.5		8.4		2.1			

4. Determination of test plans and parameter selection

4.1. Numerical experimental analysis

According to the analysis and calculation of the proposed numerical experimental plan, the crack tip is assigned to the origin (0, 0) in the construction of the coordinate system. Select 4 points of different positions in this coordinate system to represent the 4 regions near the fracture.

Select any point A near the crack tip (x=0.187 m, y=0.28 m). Figure 3 shows the temperature variation curve of point A that increases along with time under the influence of temperatures.

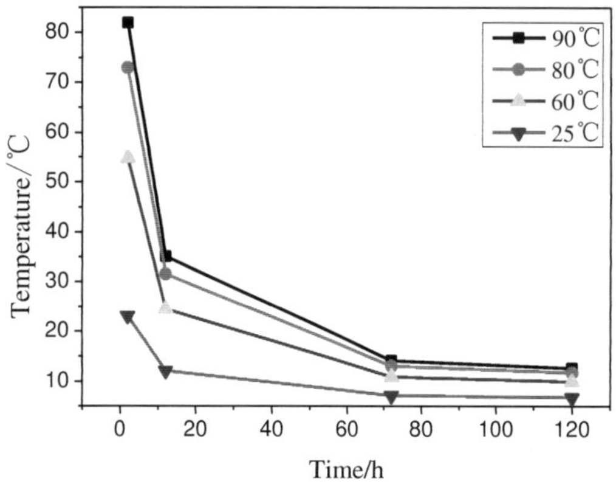

Figure 3. Temperature variation curve of Point A.

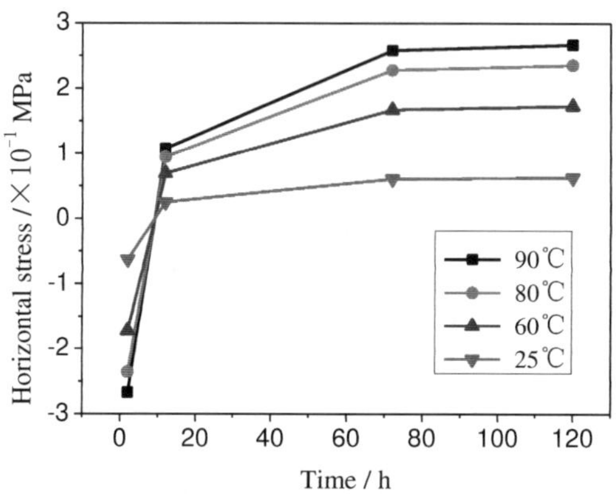

Figure 4. Horizontal stress curve of Point A.

Noting in Figure 3, for any selected determination point A, as the time passes, the variation of the temperature tends to be stable (x=0.187 m, y=-0.283 m) after 72 h. During the initial propagation (2~12 h) of the temperature, the temperature ramp of 90°C is the largest for the surrounding rocks, and that of 25°C is the smallest. As such, the temperature ramp of 90°C is 4.25 times that of 25°C; and the temperature ramp of 80°C and 60°C is 3.75 times and 2.75 times that of 25°C, respectively. The temperature ramp rises as the temperature difference increases. During the period from 12~72 h, the temperature ramp tends to decrease. For the 90°C, the temperature ramp is only 7.26% of that during the period from 2~12h. The temperature ramp of the other three temperatures almost reaches approximately 7.45% on average of that during the period from 2~12 h. Over time, the temperature difference between the fracture and the surrounding rock tends to gradually approach balance.

Figure 4 shows the horizontal stress curve of point A that increases along with time under temperatures.

The horizontal stress generated by the surrounding rock at 90°C is the largest. The stress at 90°C concentration growth rate near the fracture is 4.3 times larger than that at 25°C. Temperature difference between the surrounding rock and the fracture decreases as the temperature of the surrounding rock decreases, and the magnitude of its stress also reduces accordingly. With the transformation of temperature, the stress concentration around the fracture weakens continuously and tends to stay at a stable temperature after 72 h.

4.2. Comparison of results

Taking partial segment 2# of the water diversion tunnel as the object, the data of the on-site temperature measurement for the fixed point of the tunnel segment is recorded by selecting the horizontal length of 1 m relative to the fracture, and a vertical distance of 1 m at the fixed point S. At the same time, numerical experiments can be conducted by using the computational model with the same conditions. The results are shown in Figure 5.

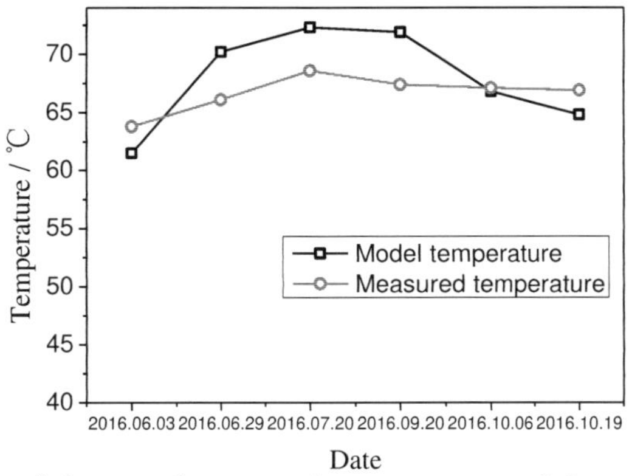

Figure 5. Comparison diagram of the actual measured temperature and the calculated temperature at point S.

It is evident from Figure 5 that the variation tends to be consistent between the actual measured temperature and the calculated temperature at point S. The maximum average temperature appeared at 1128 h of the computational model on the measurement date of 2016.07.20. The actual temperature descending ranges of 9.20~10.06 and 10.06~10.19 are 6.3 times and 5.0 times that of computational model, respectively. The actual temperature descending range of 10.06~10.19 (13 d) was reduced by 61% from 9.20~10.06 (16 d) within the same period. The reason is that the branch tunnel adjacent to the 2# tunnel segment is connected to the main tunnel after 9.20, so the air convection is strengthened through the exchange of air with heat from the external environment. However, the working condition during this period was not taken into consideration in the computational model.

5. Conclusions

The numerical experiments of the fractured rock mass at different temperatures indicate that the propagation of different temperature fields in the tunnel surrounding rock satisfies a certain fracture pattern and tends to stabilize after three days of heat transfer.

The surrounding rock stress concentration is positively correlated to the temperature difference between the high-temperature tunnel surrounding rocks. A greater temperature difference indicates greater rock instability, and when the fracture zone of the rock contains a large temperature difference, the surrounding rock is more prone to instability.

The numerical experimental results are consistent with the measured outcomes, which means that the development of the model has both a scientific and a practical value. The systematic numerical experiment can create a solid theoretical foundation for further engineering design.

Acknowledgments

The study was sponsored by the National Natural Science Foundation of China (Grant No. 51509200), the Nurturing Foundation of Xi'an University of Science and Technology (Grant No. 6310214036) and the Doctoral Foundation of Xi'an University of Science and Technology (Grant No. 6310115020).

References

[1] Simpson C 1985 *Journal of Structural Geology* **18** 503–511
[2] Karel S and Bedeiich M 1996 *Journal of Structural Geology* **18** 719–733
[3] Alsha Y.N.A, Khan K and Abduljauward S. N 2000 *International J.of Rock Mechanics and Mining Sciences* **37** 629–643
[4] Dwivedi R.D, Goel R.K, and Prasad V.V.R 2008 *A Sinha Thermo-mechanical properties of Indian and other granites* **45** 303–315
[5] Noorishad J, Tsang C. F and Witherspoon P.A 1984 *J. Geophyte Res.***89** 165
[6] Li N , Xu B, Chen X.F 2006 *China Journal of Highway and Transport* **19** 1–5
[7] Bahr H. A, Balke H and Kuna M 1987 *Theoretical Applied Fracture Mechanics* 33–39

IOP Conf. Series: Materials Science and Engineering **362** (2018) 012015 doi:10.1088/1757-899X/362/1/012015

Research on the Decision Method of Maintenance Materials Direct Supply

Qian Zhu[1], Xiaopei Shi[2], Shenyang Liu[1], Guangxu Luo[1], Chen Zhu[2]

[1]Department of Aviation Four Stations, Air Force Logistics College, Xige Street, Xuzhou, China
[2]Department of Air Materia, Air Force Logistics College, Xige Street, Xuzhou, China

Abstract. With the further development of civil military integration, more and more maintenance materials will be supplied by the factory directly. Aiming at the mode condition of maintenance materials factory direct supply, maintenance materials needs equipment support in the process of facing a number of direct supply manufacturers how to decision problems, using AHP, considering many factors optimization of direct supply manufacturers involved, and gives the weights of the evaluation indexes of the direct supply manufacturers to evaluate optimal. Finally, with 4 straights for the manufacturer as an example, considering the various evaluation indexes to carry out evaluation and drawing the correct evaluation of direct supply manufacturers, the best manufacturers direct supply is selected. An example shows that, AHP can provide scientific and theoretical basis to materials factory direct supply security.

1. Introduction

At present, in the maintenance equipment manufacturers direct supply mode, the introduction of multiple direct manufacturers bidding competition can achieve cost efficient and transparent purposes. The commonly used evaluation methods include the lowest quotation method, the lowest evaluation method, the approximate bottom price method, the experience evaluation method and the comprehensive score method [1]. The above methods will inevitably appear unreasonable and unfair because of the Jue factors, market supply and demand factors, time, geographical location and traffic conditions, and the units which choose maintenance equipment manufacturers' direct supply mode have some characteristics, such as emergency, decentralized and after-sale service requirements, which require the units with direct supply of manufacturers to consider more factors to complete the business process such as decision - making, cooperation intention, and the purpose of promoting the deep development of integration of defense and civilian technologies, the improvement of equipment availability rate, and the efficient and transparent use of cost [2]. This paper mainly discusses the application of the maintenance equipment manufacturers direct supply guarantee mode in the maintenance equipment direct suppliers to tender response, using the analytic hierarchy process, to optimize the direct supply manufacturers need to consider many factors, to evaluate the decision - making optimal bidding for manufacturers.

In the direct supply mode of maintenance equipment manufacturer, the user needs to feedback the maintenance equipment through the supply chain information network system to the front end of the supply chain [3]. The maintenance equipment is directly supplied by the manufacturer under the traction of the force user, and the maintenance equipment supply is guaranteed to the user. The direct supply mode of maintenance equipment manufacturer is not only a military activity, it needs scientific management to improve military benefits, but also a complex economic activity, and has a connection with various economic factors, with the dual nature of military and economic, and is subject to the double restriction of military law and economic law [4]. Below, we briefly introduce the five indicators of precision execution rate, rapid response ability, production line flexibility, manufacturer's economic benefits and supply price stability in the optimization of maintenance equipment direct manufacturers

2. Analytic hierarchy process

Analytic hierarchy process (AHP) decomposes the complex problem into each component, and then group these factors into an ordered hierarchical structure according to the dominant relationship. The relative importance of the factors in the hierarchy is determined by the comparative method. Finally, the overall judgment determines the order of the relative importance of the factors, namely the advantages and disadvantages of each scheme, which is the basis of the evaluation and selection scheme [5]. According to the 1-9 scale method shown in Table 1, the relative importance degree of each two evaluation indexes is determined by comparison, and then the judgment matrix is constructed [6].

Table 1. The 1-9 scale method of judging matrix

Serial number	Scale	Relative comparison
1	1	it is of equal importance
2	3	one factor is slightly more important than the othe
3	5	one factor is more important than another
4	7	one factor is much more important than another
5	9	one factor is much more more important than another
6	2、4、6、8	2, 4, 6, 8, median of the two adjacent judgments
7	Reciprocal	the inverse of the other factor is inversely proportional to the original factor

(1) Determine the solution method. The calculation method has the root method, regularization, and so on. In this paper, the root method is used to compare the maximum characteristic root of the matrix and its corresponding feature vector w, and the feature vector w normalization is the weight value of the relevant elements of a layer on the upper layer. The formula is as follows:
(2) Consistency check. In calculating the judgment matrix, in order to measure whether the judgment matrix has the quan yi property, we introduce the average random consistency index ri of the judgment matrix, and the random mean value of the consistency index of the same order matrix, whose value is shown in Table 2.

Table 2. The value of the random consistency index RI

Order	1	2	3	4	5
RI	0	0	0.57	0.89	1.11
Order	6	7	8	9	
RI	1.23	1.31	1.39	1.44	

The random consistency ratio Cr is the ratio of the consistency index ci to the average random consistency index ri of the same order, namely [7].
Determining the contact form between the matrix elements is shown in Table 3.

Table 3. Determining the contact form between the matrix elements

A	B_1	B_2	...	B_n	W
B_1	b_{11}	b_{12}	...	b_{1m}	b_1
B_2	b_{21}	b_{22}	...	b_{2m}	b_2
⋮	⋮	⋮		⋮	⋮
B_m	b_{m1}	b_{m2}	...	b_{mm}	b_m

3. Applications

The application of AHP in the direct supply mode of maintenance equipment manufacturers can effectively solve the various indicators of the advantages and disadvantages, and even the contradictions of various indicators, such as good quality but higher quotation. Therefore, in order to achieve the expected target, the equipment needs to make comprehensive evaluation of multiple direct manufacturers, and give a certain weight to each evaluation index. The following four direct manufacturers, five evaluation indicators as an example to illustrate how to apply AHP in the equipment manufacturer's direct guarantee mode.
According to the 1-9 scale method of the judgment matrix shown in table 1, the optimal direct supply manufacturers factors are compared on one item by item, and the a-b judgment matrix is shown in Table 4.

Table 4. The a-b judgment matrix

A	B_1	B_2	B_3	B_4	B_5
B_1	1	3	5	6	1/2
B_2	1/3	1	3	4	1/7
B_3	1/5	3	1	2	1/5
B_4	1/6	1/4	1/2	1	1/8
B_5	2	7	5	8	1

The values of judging matrix is shown in Table 5.

Table 5. The values of judging matrix

	B_1	B_2	B_3	B_4	B_5	
A_1	0.12	0.15	0.26	0.09	0.36	W
C_1	0.10	0.17	0.22	0.52	0.11	0.15
C_2	0.47	0.21	0.21	0.108	0.18	0.26
C_3	0.27	0.48	0.16	0.30	0.30	0.24
C_4	0.14	0.11	0.39	0.061	0.39	0.33

From Table 5, it could be found that the sort of C1, C2, C3, C4 is 4, 2, 3, 1.

4. Conclusions

In the maintenance equipment manufacturer direct supply mode, how to correctly screen out the best direct suppliers to take the direct guarantee mode of the unit to provide equipment support, to improve the comprehensive protection ability is very important. The research of evaluation system is also very important to the direct guarantee mode of the equipment manufacturers. However, it must be noted that the evaluation system of this paper has limitations, and some special interference factors have not been taken into account.

5. References

[1] SHI, QuanWANG, Li-xinSHI and Xian-mingZHAO Mei 2016 System Decision-making and Modeling Beijing. *J. National Defence Industry Press* **1** 23-24

[2] Fa-ming, WANGNing and CAI 2008 Evaluation of International Competition of Petroleum Enterprises Based on Analytic Hierarchy Process. *J, Statistics & Decision-making* **5** 181-183

[3] Tie-ning, WANGHong-wei and WANGYu CAO 2009 Research on Joint Managed Inventory Strategies of Material Suly Chain. *J. China Logistics & Purchasing* **13** 56-57

[4] R DekkerM, J KleijnP and Jde Roo 2014 A Spare Parts Stocking Policy Based on Equipment *J. Criticality Production Economic* **56** 69-77

[5] Xiao-hui CHEN, Tian-wen SHENG and Shu-ping YI 2012 The Ordering Strategy of Spare Parts of Multi-Unit System Based on Periodic Preventive Maintenance. *J. Journal of South China University of Technology* **37** 95-99

[6] ZHANG M, HXU Q and SMASSART D L 2004 Averaged and Weighted Averaged Partial *J. Least Squares Analytica Chimica Acta* **4** 279-289

[7] ZIAD Ramadan and PHILIP K 2005 Alication of PLS and Back Propagation Neural Networks For the estimation of soil properties. *J. Chemometrics and Intelligent Laboratory* **5** 23-30

IOP Conf. Series: Materials Science and Engineering **362** (2018) 012016 doi:10.1088/1757-899X/362/1/012016

Influences of the Ratio of Polyol and MDI on the Acoustic Parameters of Polyurethane

Yonghua Wang[1], Zheming Liu[1], Haiquan Wu, Chengchun Zhang[2], Huadong Yu[1], Luquan Ren[2], Mohamed Ichchou[S]

[1]Ultra precision machining laboratory, Changchun University of Science and Technology, Changchun 130021， China
[2]Key Laboratory of BionicEngineering of Ministry of Education,Jilin University, Changchun 130025, China
[3]Laboratory of Fluid Mechanics and Acoustics, Ecole Centrale de Lyon, Ecully 69131, France
E-mail: 45050524@163.com

Abstract. In this paper, the influence of different ratio of polyol and MDI on the absorption coefficient and acoustic parameters of polyurethane was studied. Ratio of 100:40 and 100:45 show the best sound absorption performance, and the change trend of transmission loss and sound absorption coefficient are opposite. The flow resistance increased with the increasing of the ratio of polyol and MDI, the greater the flow resistance, the worse the high frequency sound absorption property of the polyurethane. When the ratio of polyol and MDI keep 100:45, the minimum porosity of sample, the polyurethane porosity increase with the ratio of polyol and MDI increase.

1. Introduction

With the fleeting development of human society and economy, noise is more and more ruinous for human body and mind, which is becoming a worldwide problem. There are many methods to control the noise pollution， but the most radical method is to use sound absorbing materials.

Polyurethane, simply called PU, is a general name for the molecular structure of the polymer containing repeated amino acid ester group (-NHCOO-). PU foam, rubber, coating, adhesive, synthetic fiber, thermoplastic elastomer, biomedical materials and other products can be made in terms of various requirements [1-2], it used extensively in transportation, construction, machinery, electronic equipment, furniture, textile, petrochemical, water conservancy, national defense, medical and other fields [3-9].

In the 1970s and 80s, the modern multifunctional composite materials with PU foam as the main material was developed at domestic and aboard [10]. Such as the shock absorption, sound absorption, sound insulation composite material which developed by German Terson company, the thin and elastic PU film can be used to protect the foam from liquid medium, oil and dust damage, and has good sound absorption property. Japanese Takanisawa Cybernetics company developed a composite material which make the PU foam be the main material, make a double-layer and multilayer composite materials. This material has been used in various occasions such as mechanical shell, vehicle engine cover and sound room. English researchers use flame retardant PU foam to produce sound absorbing prefabricated parts [11], can be cut or molded is the advantage of this PU foam sound absorbing product. Recently, the Navy department of United States used PU foam composite with other materials (epoxy resin, silicone gel and carboxyl rubber), successfully developed a new type of noise barrier material, improved the ability of sound barrier 50%~100% than before. A new type 6cm thick composite material can be equal to 25.4mm thick concrete. Besides, this new type of material can be used as a sound barrier in the whole spectrum, especially have a great effect on low frequency noise.

As a kind of acoustic material, PU is usually made into porous PU foam, it really have an amazing sound absorption performance, which is a popular new type of acoustic material. Compared with the normal used superfine glass wool, rock wool, slag wool and other fibrous sound-absorbing material, PU has a series of advantages such as light quality, high sound absorption coefficient, easy processing, no dust pollution, waterproof, moisture-proof, wide adaptability[12-16]. Therefore, people pay more attention to the PU acoustic material and its absorbent products, both domestic and foreign development of this new type of sound absorbing material at present. However, the main research is focused on the influence of preparation conditions on the sound absorption performance, and less acoustic characteristic parameters which determine the sound absorption performance. From this point of view, the paper studied the influence of the preparation ratio on the acoustic parameters.

2. Materials and methods

2.1. Material selection

The PU foam was fabricated by one step foaming method using polyether polyol HEP-330N, polyemer polyol TPOP-36/28, deionized water, silicone oil, catalysts A33 and A-1, modified MDI (the content of –NCO group is 30%). The physical and chemical properties of polyols are shown in Table 1. The proportion of each material compared with the total mass of polyols (100 fractions in all) is displayed in Table 2. In this experiment, the total mass of raw material was 42g, and the temperature was 55℃, the ratio of polyol to modified MDI was 100:40, 100:45, 100:50, 100:55, 100:60, each experiment was repeated three times to reduce the error.

Table 1: Physicochemical properties of polyhydric alcohols

	Polyol330N	Polyol36/28
Hydroxyl value (mg KOH/g)	33-37	25-29
acid value (mg KOH/g)	≤0.1	≤0.05
Humidity (%)	≤0.05	≤0.05
pH	5.0-7.0	6.0-9.0
viscosity (mPa•s/25°C)	800-1000	≤3500
molecular weight	5000	4000

Table 2. Basic formulation of preparation PU

Raw material	mass fraction	Component
Polyol330N	60	Component A
Polyol36/28	40	
Deionized water	2	
silicone oil（8719）	1.8	
A33	0.6	
Two ethanolamine	0.5	
A-1	0.1	
Modified MDI	45	Component B

2.2. Preparation method

The specific preparation process of PU foam is described in Figure 1. The mould was kept in baking oven for more than 30 minutes until it reached the designed foaming temperature. The mixture was stirred for 10-15 minutes at high speed of 2000 r/min with a mechanical stirrer. Poured the mixture into the mould rapidly at the cream time and postcured for 5 minutes. For the preparation of PU-RH composites, RH was weighed 2%, 5% and 8% of the total mass of all the raw materials respectively and then mixed with A component (all the others raw materials except modified MDI) at room temperature before mixing with B component (modified MDI).

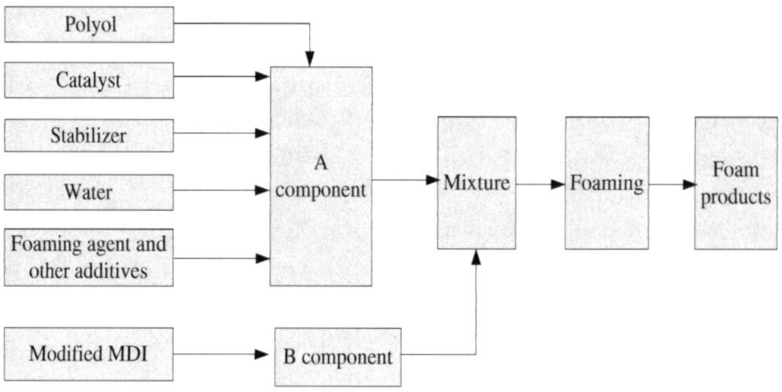

Figure 1. The preparation process of PU foam

2.3. Sound absorption and insulation test

The main methods include standing wave tube method, transmission loss method and reverberation chamber method, in this paper, the transfer function method is used to test the sound absorption coefficient of the material. The test equipment is the SW series impedance tube and four channel acoustic analyzer from Beijing reputation Acoustic Electric Technology Co., Ltd.. Four kinds of acoustic properties was tested, acoustic impedance, sound reflection coefficient, sound absorption coefficient and transmission loss. The equipment(a) used in this test, transmission loss principle(b) and testing principle of sound absorption coefficient (c) was shown in figure 2. The impedance tube diameter is 100mm, the sampling frequency of dSPACE is 51200Hz. A wide random sound wave generated by a loudspeaker at one end of the impedance tube was transmitted to the material surface of the other end.

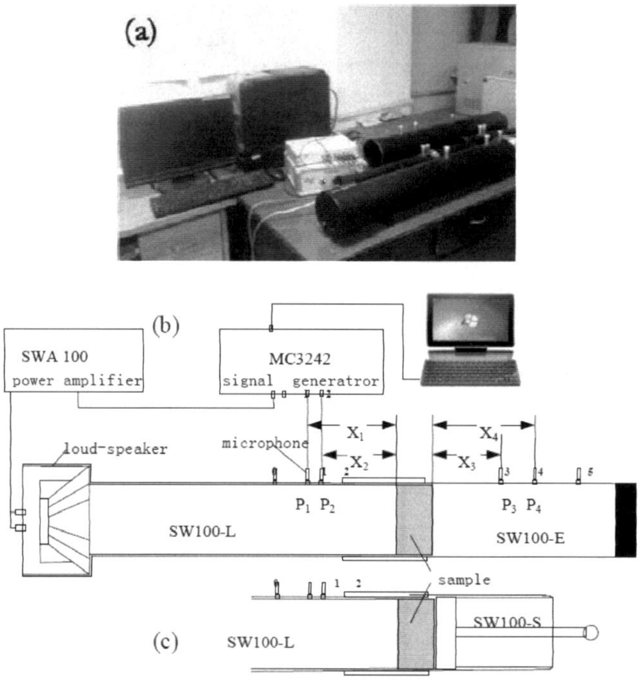

Figure 2. Impedance tube and its test principle

The transmission loss can be calculated as:

$$TL = -20\lg(\frac{\sin[k(X_1 - X_2)]}{\sin[k(X_3 - X_4)]} \times \frac{P_4 e^{jk(X_3-X_4)} - P_3}{P_1 - P_2 e^{-jk(X_1-X_2)}} \times e^{jk(X_2+X_4)})$$

The X_1, X_2, X_3, X_4 is the distance between the four microphones and the surface of the test sample respectively. The P_1, P_2, P_3, P_4 are corresponding position of complex sound pressure respectively. K is the complex wave number, Z_s is the acoustic impedance; r is the sound reflection factor; α is the sound absorption coefficient; TL is the transmission loss; ρ_0 and c_0 are the density of air and the speed of sound respectively.

The incident sound was perpendicular to the surface of the foam rise direction. Each of the tests was repeated with at least five samples to obtain consistent and representative results, and made at the temperature of (20 ± 2)°C and relative humidity of (60 ± 10)%.

2.4. Test of characteristic parameters

Using a set of experimental equipment made by myself to measure the acoustic characteristic parameters of PU. The principle and method are as follows the paper: "a method for testing of the characteristic parameters of porous materials rapidly". The main parameters of the test include flow resistance, porosity, curvature, viscous characteristic constant and thermal characteristic constant.

3. Results and discussion

3.1. Sound absorption properties of different PU

Figure 3 shows the comparison of the sound absorption coefficient of the samples prepared by different ratios of polyol and MDI. From the figure, the ratio of polyol and MDI has a significantly influence on the absorption coefficient of PU. Under the ratio of 100:40 and 100:45, the absorption performance of PU is the best, and the peak value is close to 1. More than 100:50, the sound absorption properties of PU decreased sharply, especially 100:60, the peak value was not more than 0.5. Therefore, the suitable ratio of polyol and MDI is between 100:40-100:50.

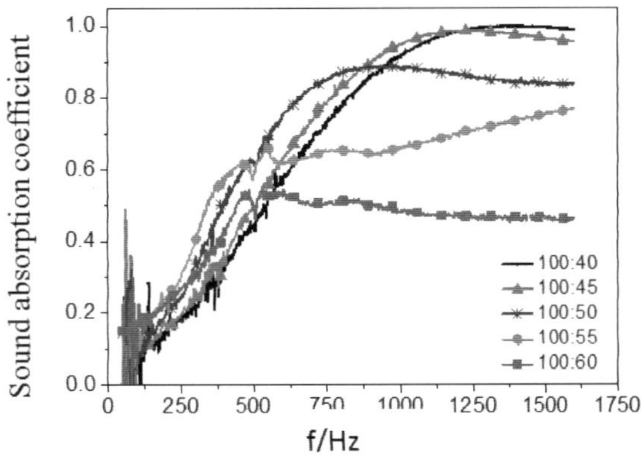

Figure 3. The absorption coefficient of PU with different ratios

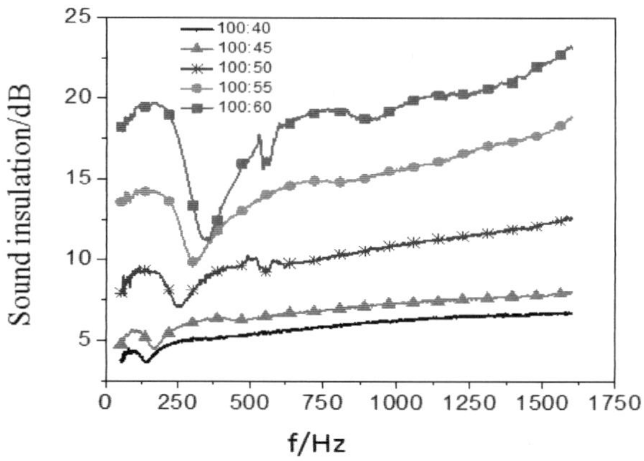

Figure 4. Transmission loss of PU with different ratios

3.2. Sound insulation performance of different PU

The contrast of the transmission loss of PU with different polyol and MDI ratio was shown in Figure 4. The figure showed that the change trend of transmission loss and sound absorption coefficient are opposite.

3.3. Acoustic parameters of different PU

(1) Sound velocity test

Test characteristic parameter of polyurethane with different ratio of polyol and MDI. Table 3 shows two methods to test the values of sound speed and air density, it shows two results of sound speed difference is relatively large, in order to make the sound speed test more accurate, ensure the repeatability of the results, use two microphone method to test.

Table 3. The sound speed and air density of PU with different ratios

Sample	Air density (g/cm3)	Sound speed（m/s method 1）	Sound speed（m/s method 2）	Sound velocity difference(m/s)
100:40	1.2100	337.0247	344.3000	7.2753
100:45	1.1944	338.9808	345.000	6.0192
100:50	1.1960	338.2439	346.9000	8.6561
100:55	1.1811	340.2290	343.4000	3.171
100:60	1.1736	341.2543	344.6000	3.3457

(2) Sound absorption performance test of self-made equipment

Figure 5 shows the sound absorption coefficient of PU with different ratio of polyol and MDI. Compare with the test results with equipment of BSWA in figure 3, the influence trend is completely consistent, sound absorption are best at ratio of 100:40 and 100:45. The sound absorption properties of PU decreased sharply when the ratio is larger than 100:50. The results show that the self-made equipment is reliable and effective.

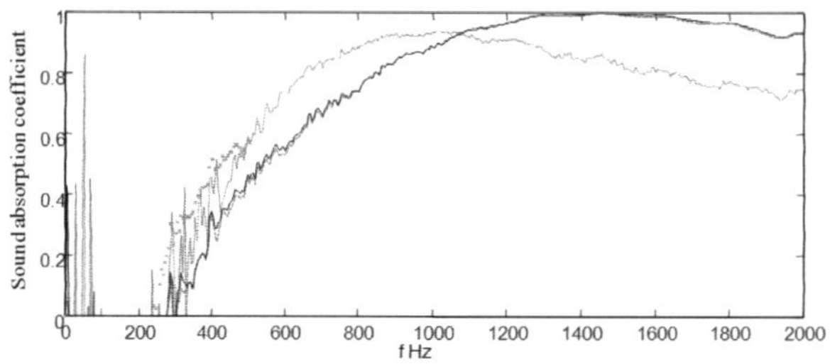

Figure 5. Comparison of sound absorption coefficient between PU with different ratios

(3) Characteristic parameter test

Characteristic parameter of each sample material obtained by MATLAB calculation are shown in table 4, the table showed that the flow resistance increases with the ratio increase, the greater the flow resistance, the worse the high frequency sound absorption property of PU. And low frequency sound absorption performance is better, but the change of the high frequency sound absorption coefficient is larger, that the best flow resistance of PU porous material is about 1000 Pa • s, the porosity is relatively stable, more than 0.9.

Table 4. Characteristic parameters of PU with different ratio of polyol and MDI

sample	Flow resistance(rayls/m)	porosity	curvature	Viscous shape factor	Thermal shape factor	M'
100:40	10988	0.9251	1	0.2992	1.7451	2.1374
100:45	9682.2	0.9106	1	0.3018	2.0122	1.0007
100:50	18097	0.9413	1.4	0.2455	2.7639	1.0024
100:55	39446	0.9843	1.1	0.2959	1.9234	1.0002
100:60	45279	0.99613	1.5	0.3428	2.3921	1.0013

The measured flow resistance and porosity of the samples are shown in Figure 6, the change of flow resistance with the ratio of polyol and MDI in the first decreased and then increased. Minimum sample flow resistance when the ratio of polyol and MDI keep 100:45, and then with the increasing of proportion, the flow resistance of the samples showed an upward trend. The trend of porosity is the same as flow resistance, minimum porosity of sample when The ratio of polyol and MDI keep 100:45, minimum porosity of sample, and then the proportion increased, PU porosity increasing. The increase of porosity is due to the reaction of polyol and MDI more completely, create more small and dense holes.

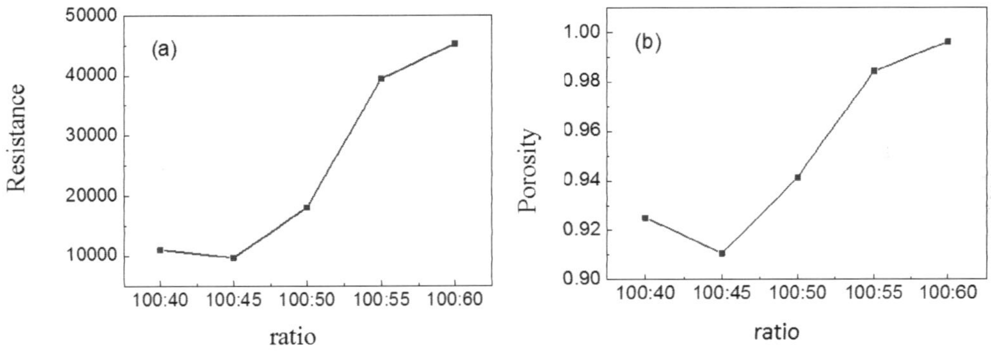

Figure 6. The effect of the ratio of polyol and MDI on the flow resistance (a) and porosity (b) of PU

(4) Comparison of predicted and experimental values

In condition of 30mm of back cavity, predict the sound absorption coefficient of samples with different ratios of polyol and MDI, compare with the obtain values and the test values, the results are shown in Figure 7 ,the predicted values are in almost same with the experimental values in addition to 100:55, especially in the middle and high frequency of 500Hz, maximum less than 5%. For the 100:55 samples, the predictive value and the test value have a large difference, some frequencies node even reach 16%. This is due to an excess of isocyanate, the formation of the foam inner hole is different from the ideal, make the test curve of the sound absorption coefficient is not smooth enough, and predict use the Johnson-Allard model and Lafarge-Allard model will derive a huge error.

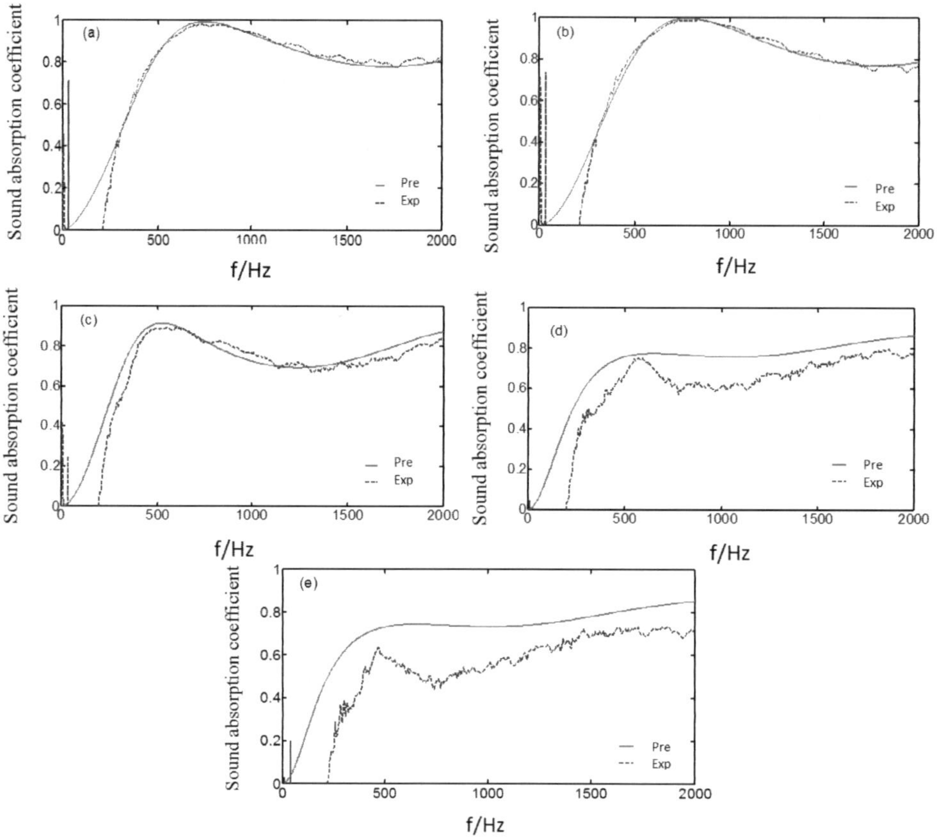

Figure 7. Comparison of the predicted and experimental values of the sound absorption coefficient of PU with different ratios

4. Conclusion

The study prepared kinds of PU with different ratio of polyol and MDI, tested the effect of the ratio on sound absorption and acoustic characteristic parameters of PU foam. Compare of the test results, although the impedance tube cross-section is different, the sound absorption coefficient curve of the same sample is different, but the proportion of polyol and MDI on sound absorption properties of PU trend is completely consistent. The proportion of polyol and MDI had significant influence on the sound absorption performance of PU. Analyze the effects of ratio of polyol and MDI of the flow resistance and porosity of PU foam, the characteristic parameters were obtained through Johnson-Allard model and Lafarge-Allard model, the sound absorption coefficient of the sample backed with 30mm cavity is compared with the test values, except for a few samples, the predicted values are highly consistent with the test values, especially in the middle and high frequency of 500Hz.

5. Acknowledgments

All support is greatly acknowledged and appreciated, especially the constructive discussion and criticism from colleagues. Thanks are due to Project supported by the National Natural Science Foundation of China(Grant No. 51505039), The Education Department of Jilin Province, "5th thirteen-Year" scientific research project planning(Grant No.KYC-JC-XM-2016-036) and LuQuan Innovation Fund Project(bionic porous sound absorption structure which based on multi population genetic algorithm be optimal design by demand) provide financial support for this article.

6. References

[1] Howard G T 2002 Biodegradation of polyurethane: a review *J. International Biodeterioration & Biodegradation.***4** 245-252

[2] Janik H, Marzec M, 2015. A review: fabrication of porous polyurethane scaffolds *J. Materials Science & Engineering C Materials for Biological Alications.* **48** 586-591

[3] Tsai H C, Hong P D and Yen M S 2007 Preparation and Physical Properties of MDEA-based Polyurethane Cationomers and their Alication to Textile Coatings *J. Textile Research Journal.* **9** 710-720

[4] Karger-Kocsis J, Gremmels J and Mousa A 2000 Alication of hygrothermally decomposed polyurethane in rubber recipes: Part 1: Natural rubber (NR) and nitrile rubber (NBR) stocks *J. KGK Kautschuk Gummi Kunststoffe.* **9** 528-533

[5] Yari A, Yeganeh H and Bakhshi H 2014 Preparation and characterization of novel antibacterial castor oil-based polyurethane membranes for wound dressing alication *J. Journal of Biomedical Materials Research Part A.* **102** 84-96

[6] Li J, Zheng W and Zeng W 2014 Structure, properties and alication of a novel low-glossed waterborne polyurethane *J. Alied Surface Science.* **18** 255-262

[7] Fang C, Zhou X and Yu Q 2014 Synthesis and characterization of low crystalline waterborne polyurethane for potential alication in water-based ink binder *J. Progress in Organic Coatings* **1** 61–71

[8] Cho YS, Lee J W and Lee J S 2002 Hyaluronic acid and silver sulfadiazine-impregnated polyurethane foams for wound dressing alication *J. Journal of Materials Science: Materials in Medicine.* **9** 861-865

[9] Dmitrienko S G, Pyatkova L N and Zolotov Y A 2002 Sorption of Ion Associates on Polyurethane Foams and Its Alication to Sorption–Spectroscopic and Test Methods of Analysis *J. Journal of Analytical Chemistry.* **10** 875-881

[10] Xu S, Chen L and Gong M 2016 Characterization and engineering alication of a novel ceramic composite insulation material *J. Composites Part B Engineering.* **111** 143-147

[11] RuiQi Xiang 1986 The acoustical material (or structure) and its alication of UK *J. Noise and Vibration Control* **5** 17-22

[12] Hilyard N C and Cunningham A 1995 Fundamental aspects of the behaviour of automotive noise control composites *J.Progress in rubber and plastics technology.* **1** 34-55

[13] Beckmann S and Herzog M 2015 Various types of polyurethanes in the process of Chemical Recycling *J. Wissen schaftliche Beiträge.* **9** 21-22

[14] HANDA Y P and Zhang Z 2000 Layered foams: A new class of cellular polymers *J. Cellular polymers.* **4** 241-255

[15] Sahraoui S, Mariez E and Etchessahar M 2000 Mechanical testing of polymeric foams at low frequency *J. Polymer testing.* **20** 93-96

[16] Yang T L, Chiang D M and Chen R 2001 Development of a novel porous laminated composite material for high sound absorption *J. Journal of Vibration and Control.* **5** 675-698

IOP Conf. Series: Materials Science and Engineering **362** (2018) 012017 doi:10.1088/1757-899X/362/1/012017

The Calculation of VOCs Diffusion Coefficient for Building Materials

Xin Zhang[1], Quancai Deng[1], Haijiang Chen[2] and Xiaoyun Wu[1]

[1]Hebei Institute of Architecture and Civil Engineering , ZhangjiaKou, 075024, China
[2]China coal Zhangjiakou Coal Mine Machinery limited-liability company, ZhangjiaKou, 075025, China
E-mail: mpm200064@sina.com.cn,419907484@qq.com, chj2099@126.com, wuxiaoyun_03@163.com

Abstract. Volatile Organic Compounds (VOCS), as one of the major sources of air contaminations, has an important bearing on one's general health. The adsorption capacity and velocity of the material for VOCs can be described separately using. In this paper, the detailed process and method of VOCs diffusion and partition coefficients by genetic algorithm is introduced, the algorithm is realized easily by computer program and the result by the method is precise and practical.

1. Introduction

In recent years, with the development of economy, the extensive fitting-up craze has appeared in the city and the indoor pollutant problem is outstanding day by day. The report found about hundreds of thousands of people are killed every year from decorate pollution in our country, the experts say the killer is VOCs. Because of the greatest threat to people, many experts are studying the diffusion problem of VOCs in building materials [1-3]. Among them, the Bodalal model was cited by many scholars, which described VOCs diffusion procedure by partial differential equation. Genetic Algorithms (GAs) are search algorithms based on of natural evolution processing including selection, mutation and crossover operations on the genes of individuals or potential solutions. and adopts a natural evolution mechanism to represent complex phenomenon. It searches through the total solution space and can find the optimal solution globally over a domain. In this paper, the calculation of VOCs diffusion and partition coefficients by genetic algorithm is introduced for the Bodalal model.

2. Introduction of The Model

Bodalal segregated a large chamber into two equal volumes of chambers. A certain concentration of VOC is introduced into one of the two equal volumes of chambers. After a period of time, Bodalal measured variation of VOC concentration in two chambers, and then the mathematical model of the diffusion process been established.

The model can be described as follows:

The diffusion equation is:

$$\frac{\partial C}{\partial t} = D\frac{\partial^2 C}{\partial x^2} \tag{1}$$

D is diffusion coefficient of the compound within the specimen;

C is concentration of the compound within the specimen;

x is linear distance
t is time.
Boundary conditions
At $x = 0, C = k_e C_1$
At $x = l, C = k_e C_2$
Initial condition
At $t = 0, C = 0, 0 \le x \le l$

The concentration in the low and high concentration compartments are expressed by Eqns. (2) and (3), respectively

$$C_1 = \frac{DA}{V_1}\int_0^t (\frac{\partial C}{\partial x})_{x=0} dt \tag{2}$$

$$C_2 = C_0 - \frac{DA}{V_2}\int_0^t (\frac{\partial C}{\partial x})_{x=l} dt \tag{3}$$

Eventually, the equations are derived

$$\ln\frac{(C_2 - C_1)}{C_0} = -\left(\frac{q^2 D}{l^2}\right)t + \ln\left(\frac{4\alpha}{q^2 + \alpha(2+\alpha)}\right) \tag{4}$$

$$\tan(q) = \frac{2\alpha q}{(q^2 - \alpha^2)} \tag{5}$$

$$\alpha = \frac{k_e A l}{V} \tag{6}$$

Called

$$r = \frac{-q^2 D}{l^2}, \quad u = ln\left(\frac{4\alpha}{q^2 + \alpha(2+\alpha)}\right).$$

Where q are the positive roots of the characteristic equation:

$$\tan q = \frac{2\alpha q}{q^2 - \alpha^2}$$

3. The Calculation of The Diffusion Coefficient

3.1. Graphical method

The Bodalal model used graphical method to calculate the VOCs diffusion and partition coefficients, in reference [4]. Details are as follows:

Firstly, experimentally determined values C_0, C_1, C_2, t, According to the formula, the straight line of the plot $\ln\frac{C_2 - C_1}{C_0}$ vs. t is drawn, the values of r and u are calculated by the least-square regression method.

Secondly, the picture of $\tan q = \frac{2\alpha q}{q^2 - \alpha^2}$ is drawn, where the range of α is 0 to 3. Under the same coordinate system, the curve of q vs. α is drawn with different values of u as shown in Figure1.

IOP Conf. Series: Materials Science and Engineering **362** (2018) 012017 doi:10.1088/1757-899X/362/1/012017

Finally, by observing the intersections of curve 1 and 2, the values of α and q are concluded. the diffusion and partition coefficients are calculated using $r = \frac{-q^2 D}{l^2}$ and $\alpha = \frac{k_e Al}{V}$, respectively.

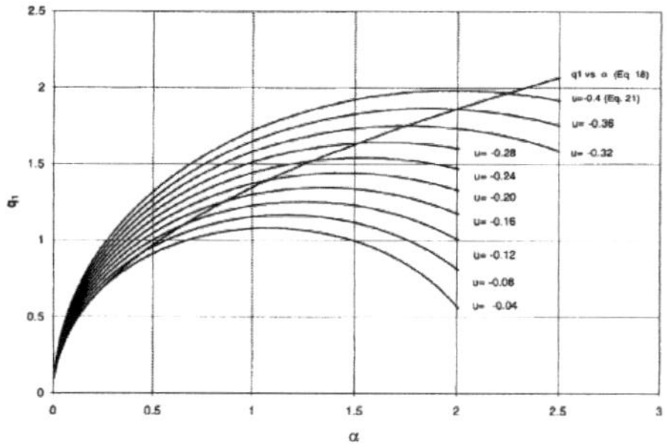

Figure 1. q vs. α

It is not difficult to find that the drawing method has the following disadvantages:

1. The solution by Drawing is not accurate, so there will be a great error. when $u = -0.32$,the values of α and q as shown in Figure1 are 1.75 and 1.76. Substituting α and q into the equation

$$\tan q = \frac{2\alpha q}{q^2 - \alpha^2}$$

After close examination, it is not difficult to draw that the value of the left side of the equation is -5.222, but the value of the right side of the equation is -175.498. The value is not the same on both sides, it is obviously wrong.

2. The range of u is only -0.4 to -0.04 as shown in Figure1, the values of u are limited which cannot meet the practical demands.

3.2. Genetic Algorithms

In view of the problem of the graphic method, in reference [5], based on the function extremum theory, Extremum-Search Method is designed by the author. The algorithm is effective to solve the problems of the graphical method and can be realized easily by the computer program, but step size of Extremum-Search Method is certain to search, it has low probability of finding the global optimal solution efficiency and tend to be trapped by local optimizations.

In the Bodalal model, the solving process of and with the characteristic equation

$$\tan q = \frac{2\alpha q}{q^2 - \alpha^2}$$

and the equation

$$u = ln\left(\frac{4\alpha}{q^2 + \alpha(2+\alpha)}\right)$$

which may be transformed into solving the value of α and q to make the equation was established.

$$[\tan(q)(q^2 - \alpha^2) - 2\alpha q]^2 +$$
$$\{e^u[q^2 + \alpha(2+a)] - 4\alpha\}^2 = 0$$

The above process is exactly to make the following function getting minimum value by solving the value of α and q

$$y = f(\alpha, q) = [\tan(q)(q^2 - \alpha^2) - 2\alpha q]^2 + \{e^u[q^2 + \alpha(2+a)] - 4\alpha\}^2 \quad (7)$$

It is a typical dualistic function extremum problem, and genetic algorithm (GA) can be designed to solve it. Because of the powerful operation ability of MATLAB, genetic algorithm Toolbox of MATLAB can realize the global optimization of traditional genetic algorithm, and has high accuracy, so, the extremum problem is solved with MATLAB toolbox.

The detailed process of GA is described as follows:

Firstly, the fitness function of genetic algorithm should be created, ga function in genetic algorithm and direct search toolbox is to solve the minimum of objective function, minimizing the problem's objective function, the objective function can be make into fitness function. So, the fitness function of this extremum problem is function (7).

Secondly, open the GUI interface of the genetic algorithm to set parameters. The parameters which include population size, crossover Fraction, Migration Fraction and Generations, adopt the default parameters of the Matlab toolbox, you can adjust these parameters according to the needs of your application. The generation of initial population is random.

Finally, click the Start button. GUI will automatically run the genetic algorithm program, and the run results are indicated.

Mean fitness, best fitness and current best individual can be obtained through the GUI running genetic algorithm As shown in figure 2, 3.

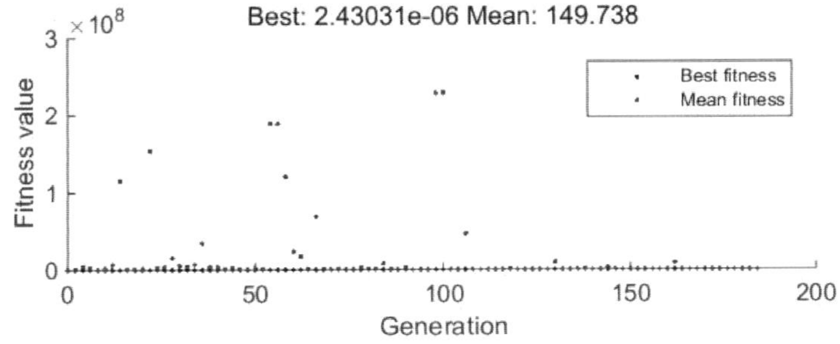

Figure 2. =0.4040, Mean fitness and best fitness

The parameters of Genetic algorithm are encoded, no apriori knowledge of the system is required, it does not has much mathematical requirements to the optimization of multiple solution problem and can improve the convergence speed and global searching ability. this could cause the search to jump out local optimization. For a given value of, the values of and can be precisely calculated. Here are several sets of data such as table 1.

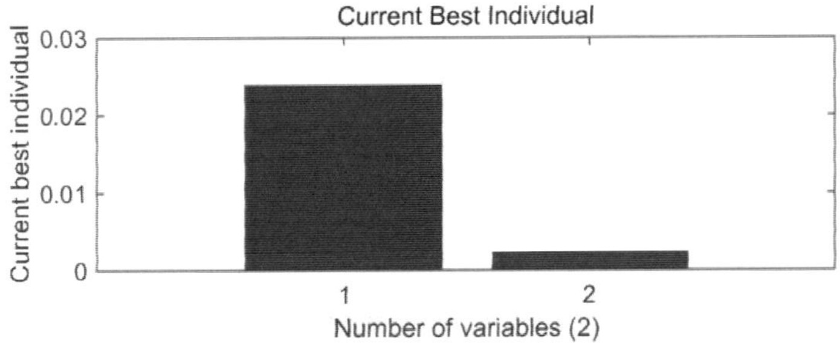

Figure 3. =0.4040, current best individual

Table 1. The data are obtained by Genetic algorithm

u	α	q	Current iteration	Best fitness
-0.4949	0.012	0.139	117	4.466*10-4
-0.122	0.06	0.376	102	1.180*10-4
-0.0458	0.216	0.665	110	7.297*10-4
-0.0288	0.007	0.038	143	1.547*10-4
-0.2666	0.062	0.416	179	8.283*10-4
0.4484	0.089	0.216	108	9.516*10-4
0.2724	0.035	0.156	90	2.615*10-4
0.404	0.002	0.024	185	2.430*10-6
0.1801	0.029	0.222	113	1.916*10-4
0.299	0.033	0.167	193	5.569*10-5

The values of and have been achieved by least square method from collecting experiment repeated data. Then according to the different values of, the values of and can be calculated.

Finally, the values of r, α and q are introduced into $r = \frac{-q^2 D}{l^2}$ and $\alpha = \frac{k_e A l}{V}$, the VOCs diffusion and partition coefficients can be obtained from the above process. Such as table 2.

Table 2. The final data

u	r	α	q	D	k_e
-0.4949	-0.3173	0.012	0.139	2.364*10-5	10
-0.122	-0.1128	0.06	0.376	1.148*10-6	50
-0.0458	-0.1076	0.216	0.665	3.503*10-7	180
-0.0288	-0.1041	0.007	0.038	1.038*10-4	5.8333
-0.2666	-0.1663	0.062	0.416	1.383*10-6	51.667
0.4484	-0.264	0.089	0.216	8.148*10-6	74.167
0.2724	-0.1983	0.035	0.156	1.173*10-5	29.167
0.404	-0.096	0.002	0.024	2.400*10-4	1.6667
0.1801	-0.0878	0.029	0.222	2.565*10-6	24.167
0.299	-0.3011	0.033	0.167	1.554*10-5	27.5

4. Conclusions

For the partial differential equation model of VOCs physicochemical evaporation process. for building materials by Bodalal. This paper introduces the genetic algorithm of VOCs diffusion and partition coefficients in dry building materials. The experiment proved that it is easy to realize the method by computer programming, the algorithm can avoid trap local optimum and has great flexibility. So it can be extended to other similar problems.

5. References

[1] Xu Y and Zhang Y P 2003 An improved mass transfer based model for analyzing VOC emissions from building materials *J. Atmospheric Environment.* **37** 2497-2505

[2] Deng BQ and Kim NC 2004 An analytical model for VOC emission from dry building materials *J. Atmospheric Environment.* **38** 1173-1180

[3] Xiong JY, Liu C and Zhang YP 2012 A general analytical model for formaldehyde and VOC emission/sorption in single-layer building materials and its alication in determining the characteristic parameters *J. Atmospheric Environment.* **47** 288-294

[4] Bodalal A, Zhang JS and Plett EG 2000 A method for measuring internal diffusion and equilibrium partition coefficients of volatile organic compounds for building materials *J. Building and Environment.* **35** 101-110

[5] Zhang X 2017 Calculation of VOCS Diffusion Coefficient Based on Extremum-search Method *J. Journal of Software Engineering.* **1** 66-71

IOP Conf. Series: Materials Science and Engineering **362** (2018) 012018 doi:10.1088/1757-899X/362/1/012018

Energy Evolution Mechanism and Confining Pressure Effect of Granite under Triaxial Loading-Unloading Cycles

WANG Hao[1,2], MIAO Sheng-jun[1,2]

[1]School of civil and environmental engineering, University of Science and Technology Beijing, Beijing 100083, China
[2]Key Laboratory of High-Efficient Mining and Safety of Metal Mines(Ministry of Education of China
E-mail: wanghao@126.com

Abstract. Rock mass undergoes some deformational failure under the action of external loads, a process known to be associated with energy dissipation and release. A triaxial loading-unloading cycle test was conducted on granite in order to investigate the energy evolution pattern of rock mass under the action of external loads. The study results demonstrated: (1) The stress peaks increased by 50% and 22% respectively and the pre-peak weakening became more apparent in the ascending process of the confining pressure from 10MPa to 30MPa; the area enclosed by the hysteresis loop corresponding to 30MPa diminished by nearly 60% than that corresponding to 10MPa, indicating a higher confining pressure prohibits rock mass from plastic deformation and shifts strain toward elastic deformation. (2) In the vicinity of the strength limit, the slope of dissipation energy increased to 1.6 from the original 0.7 and the dissipation energy grew at an accelerating rate, demonstrating stronger propagation and convergence of internal cracks. (3) At a pressure of 70% of the stress peak, the elastic energy of the granite accounted for 88% of its peak value, suggesting the rock mechanical energy from the outside mostly changes into the elastic energy inside the rock, with little energy loss.(4) Prior to test specimen failure, the axial bearing capacity dropped with a decreasing confining pressure in an essentially linear way, and the existence of confirming pressure played a role in stabilizing the axial bearing capacity.

1. Introduction

During the excavation process, rock mass undergoes some deformational failure to be associated with energy evolution. Failure of rock mass has become an important hot issue in geotechnical engineering field. The nature of rock material failure is destabilization rupture driven by energy. Now in the geotechnical engineering field, more and more people analyze the failure process of rock from the perspective of energy [1]. The failure process of rock is usually coupled with energy input, accumulation, dissipation and release, attributing to complex energy exchange [2]. Energy of rock is input into the rock in the form of external mechanical energy and is stored in rock in the form of elastic energy [2-3]. In the vicinity of the strength limit under the action of external loads, the elastic energy is released so as to generate kinetic energy, followed by causing rock fragmentation. Since it is difficult to see the micro-variation inside rock during failure process, to analyze the process from the perspective of energy is very effective and closer to the nature reason of failure [1][4][5]. Many experts and scholars have conducted some researches on the characteristics of rock failure. Zhang et al. pointed that the energy evolution process of rock reflects the failure process in the microcosmic, revealing the failure law of rocks [6]. Zhang et al. pointed that within the quasi-state loading range, the less the loading rate is, the larger the dissipation energy will be [7]. Basically, the loading rate has no influence on the elastic energy evolution. Peng et al. showed that under low confining pressure and uniaxial compression, the elasticity modulus of coal rock rises with the decrease in cyclic stress [8]. The elasticity modulus does not present such feature under high confining pressure. Yang et al., You et al. have conducted regular triaxial compression test on marble and siltstone, finding that the energy absorbed by rock mass presents the same linear relation with confining pressure [9-10].

Fu et al. studied the rock failure process under different stress paths in virtue of acoustic emission monitoring technology [11]. Based on the triaxial compression test on granite, Pan et al. found out that with the decrease in axial stress, crack inside samples increases in a linear way [12]. Zhao et al. pointed that the decrease in rock strength is due to energy dissipation which is easy to cause rock catastrophe phenomenon [13].

The excavation process of rock mass is usually coupled with energy input, accumulation, dissipation and release. The deformational failure and reinforced supporting process of loaded rock is also a process where energy transfers and evolves. Considering differences in energy input way and rate, rock structure, geological environment and other factors, rock energy evolution has different expression forms. Also, ways of rock deformational failure change accordingly. In order to observe, explain and solve some puzzles in rock engineering, it is necessary to firstly study the energy status of rock under different stress statuses, including the amount of accumulated elastic energy and dissipation energy and their proportion in total input energy. Besides, it is necessary to study the characteristics of energy evolution of loaded rock with different lithology and under different confining pressures. On the basis of triaxial cyclic loading and unloading test on three varieties of rocks respectively, the paper gets the cyclic loading and unloading curve, energy evolution and distribution law as well as the confining pressure effect. Research results are of great significance to studying the failure mechanism of rock.

2. Energy calculating method

The deformational failure process of rock is usually associated with energy dissipation and release. Considering that the release of dissipation energy features irreversibility and that of elastic energy features reversibility, we have investigated the elastic energy and dissipation energy of rock during deformational failure process. According to the Energy Conservation Law [14-15]:

$$W=Ee+Ed \tag{1}$$

Where: W—energy input from the outside; Ee—elastic energy stored inside the rock; Ed—dissipation energy during loading process.

Since elastic energy features reversibility, it is learned that in the stress-strain curve, energy corresponding to unloading springback is the elastic energy accumulated inside rock. Energy lost when there is no loading springback is dissipation energy. Figure 1 shows the loading and unloading stress-strain curve. At the stress of σ′, the dissipation energy density at the direction of stress is uid, which can be confirmed by the area between the loading and unloading curves; the elastic energy density is uie, confirmed by the area enclosed by the loading curve and the ordinate of σ′.

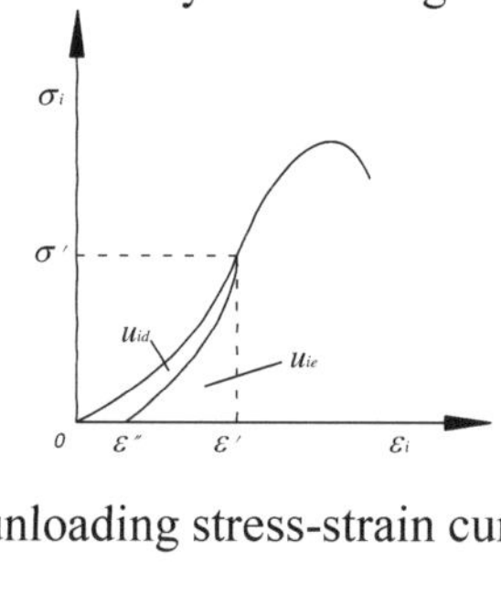

Figure 1. Loading and unloading stress-strain curves of rock at stress of σ’

3. Test process

3.1. Sample preparation

Sample is granite which is from the Lingbao Mine of Henan Province. The average longitudinal wave velocity of samples is 4,015m/s, the porosity is 1.19%, the dynamic elasticity modulus is 43.60GPa and the average density is 2.561g/cm3. There are many original interstices inside samples. Samples are processed into Φ50mm×100mm standard cylinder in lab. Two ends of samples are repeatedly polished using sandpaper to make the surface flatness lower than 0.05mm. The flatness between top and bottom surface is controlled to be lower than 0.2mm.

3.2. Test method and equipment

Choose the granite samples for triaxial loading and unloading text at confining pressure of 10MPa, 20MPa, 30MPa respectively. Use TYP-500 triaxial compression tester to add the confining pressure to target value and than add axial pressure. The confining pressure of each sample is loaded to 50% of peak limit and unloaded to 0MPa, followed by being increased by 30MPa for each time. The loading rate in early stage is 0.1mm/min. When the confining pressure is loaded to residual phase, the loading rate is rose to 0.03mm/min in order to get the post-peak curve. PCI-2 acoustic emission monitoring system is used during whole process to monitor and record data needed.

4. Rock energy evolution and analysis under triaxial loading and unloading cycle test

4.1. Triaxial cyclic loading and unloading failure characteristics

The cyclic loading and unloading stress-strain curve and failure modes of samples are got after test. As shown in Figure4, the external envelope mode in the cyclic loading and unloading stress-strain curve, similar with regular loading curve, can be divided into four phases, namely compaction phase, elastic phase, weakening phase and failure phase. The stress peaks of granite are 120MPa, 180MPa and 220MPa respectively at confining pressures of 10MPa, 20MPa and 30MPa. Therefore, when confining pressure increases, the stress peaks increase by 50% and 22% respectively. According to the stress-strain curve under the same confining pressure, increase in stress and strain presents nonlinear relation. The curve appears concave, indicating that the larger the strain is, the faster the increase speed of stress will be. According to Figure 2, 3, 4, hysteresis loop of sample deformation curve under cyclic loading effect shrinks gradually. The area enclosed by the hysteresis loop corresponding to 30MPa diminished by nearly 60% than that corresponding to 10MPa, indicating that a higher confining pressure prohibits rock mass from plastic deformation and shifts strain toward elastic deformation. Besides, prior to conducting axial stress, higher confining pressure makes natural interspace close more tight inside rock, which verifies that the higher confining pressure can improve compressive strength. After collecting and comparing fracted samples tested in lab, it is founded that there are more rock fragments under 10MPa and most fragments concentrate around the main fault line. Fault line of samples under 20MPa appears obvious shear plane and the fracture plane is not smooth, with loose rock seen. Fault line of rock samples under 30MPa is not obvious and presents curve shape. With increase in confining pressure, main fault line number decreases gradually, showing that confining pressure increase enhances the integrity of rock.

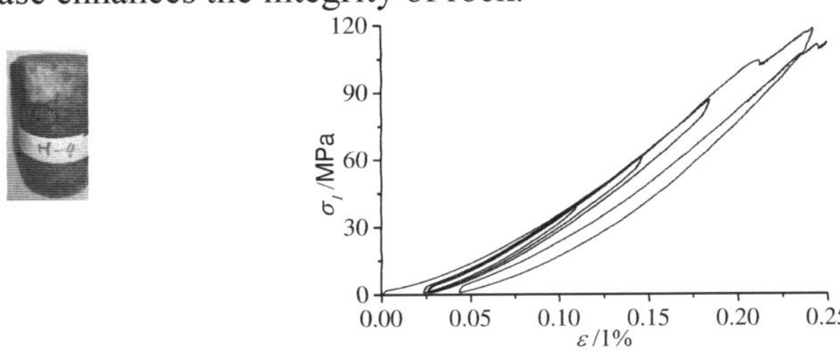

Figure 2. The cyclic loading and unloading stress-strain curves and failure modes of granite H-4 under 10MPa confining pressure

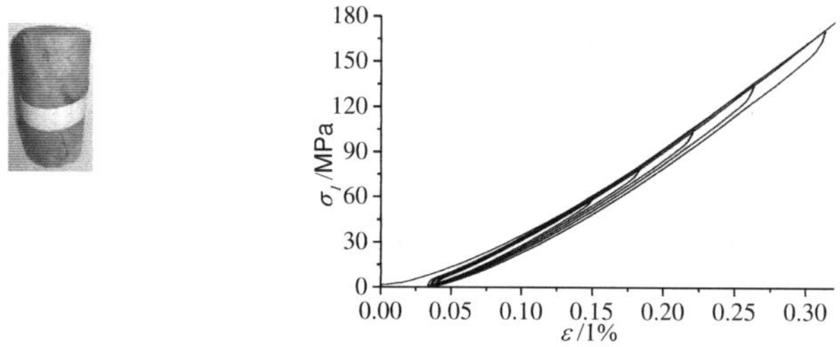

Figure 3. The cyclic loading and unloading stress-strain curves and failure modes of granite H-42 under 20MPa confining pressure

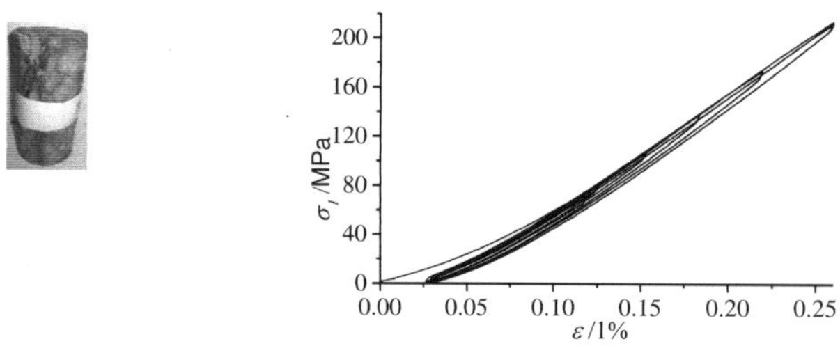

Figure 4. The cyclic loading and unloading stress-strain curves and failure modes of granite H-59 under 30MPa confining pressure

4.2. Energy evolution pattern during triaxial cyclic loading and unloading process

According to loading and unloading curve from Figure2 to Figure4, the elastic energy density and dissipation energy density of granite under different loading levels are calculated using mathematics integral formula, seen in Figure5.

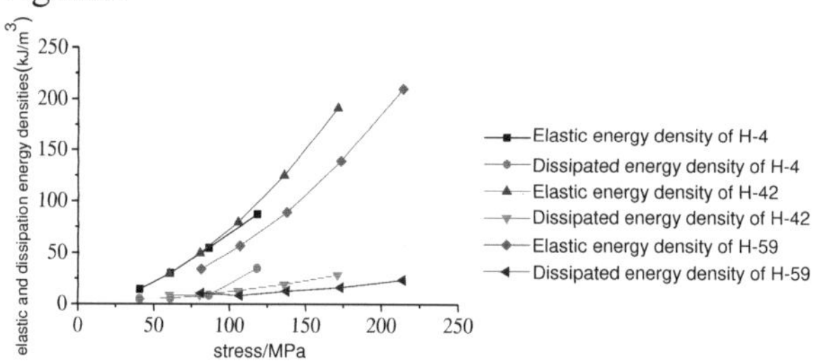

Figure 5. The elastic and dissipation energy densities curves of granite with its stress change

Theoretically, the confining pressure of hydrostatic stress status is the start point of research. Actually, the energy inside rock mass suffers some changes during the process from the beginning of loading to hydrostatic balance status, yet, the value of these changes is small compared with energy change during rock failure process. Therefore, the energy change during this phase is excluded from consideration in the test. The test only takes into account the axial action of tester on samples.

Figure5 shows that during the triaxial cyclic loading and unloading process, with the increase in axial stress, the elastic energy density curve of three samples is basically similar with dissipation energy density curve, analyzed taking H-42 as an example. The elastic energy density of sample under confining pressure of 100MPa is 85KJ/m3, and that rises to 130 KJ/m3 under confining pressure of 140MPa and to 190 KJ/m3 under confining pressure of 175MPa. As seen from image that dissipation energy curve increase is basically the same as elastic energy curve, with both increasing in an approximate power exponent form. In Figure5, since the axial stress is implemented, the elastic energy curve is basically above the elastic energy curve. Differences between two numerical values become larger and larger prior to stress peak, indicating that rock is mainly under energy accumulation prior to stress peak. From the other perspective, the reason that curve increases in an approximate power exponent form and curvature appears small to big is that there are too many interspace cracks inside rock in early stage. Inside particles are squeezed mutually after compaction and energy conversion ratio increases. With increase in loading, samples gradually reach the energy storage strength limit. In the vicinity of the strength limit, the slope of dissipation energy increased to 1.6 from the original 0.7, rising by 128% or so. The growth rate keeps rising, demonstrating stronger propagation and convergence of internal cracks. Also, the inside structure is under change under action of cyclic loads. As shown in diagram, the dissipation energy density of sample H-4 suddenly boosts in the fourth cycle because that a large amount of energy is consumed due to pre-peak failure, which is basically consistent with the research results of Dai Bing.

4.3. Energy distribution law during triaxial cyclic loading and unloading process

Figure 6 is the curve of proportion of elastic energy and dissipation energy of granite in total input energy under different confining pressures with stress peak intensity ratio change. The curve reflects the energy distribution law during triaxial cyclic loading and unloading process. During the triaxial cyclic loading and unloading process, a part of mechanical energy implemented by machine is released in the form of dissipation energy and another part is transferred into elastic energy to be stored inside the rock. In energy ratio curve, elastic energy takes up 0.7-0.9, much higher than that of dissipation energy, demonstrating that most mechanical energy from the outside are transferred into elastic energy and only 10%-30% of mechanical energy are transferred into other dissipation energy. In the ascending process of peak intensity ratio from 0.2 to 0.7, the proportion of elastic energy of H-42 in input energy rises with the increase in loads and, at the peak intensity of 0.7, the proportion reaches the peak of 0.88 while proportion of dissipation energy reaches the lowest point of 0.12 (H-59 presents similar image). That indicates that prior to stress peak, stress change impacts energy evolution process, revealing the distribution law of energy inside rock. According to the curve of H-4 granite, the proportion of elastic energy of the fourth cyclic rock lowers greatly and that of dissipation energy rises greatly because that the samples have lost bearing capacity due to failure before the fourth cycles, demonstrating that large amount of energy will be consumed during failure process which takes up a large proportion of energy consumption. Based on above analysis, value of confining pressure can obviously affect the energy distribution inside rock. The larger the confining pressure is, the larger the elastic energy it transfers will be and the easier that rock burst will occur during excavation.

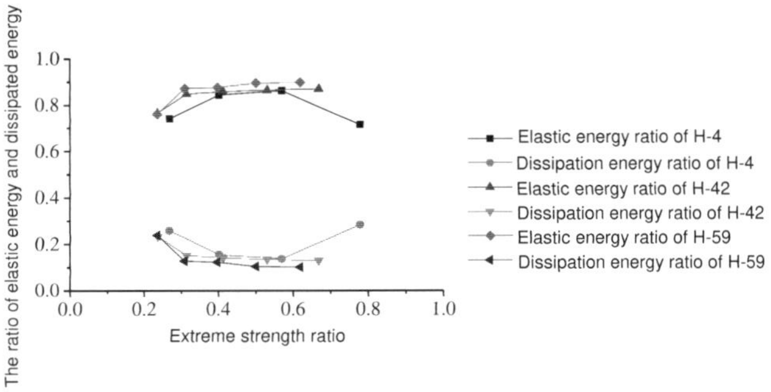

Figure 6. The evolution law of granite elastic and dissipation energy account for the proportion of total input energy with the ratio of peak intensity change

5. Confining pressure effect of rock energy evolution

5.1. Rock energy evolution under unloading confining pressure test

According to the rock inside energy evolution and distribution law got from the triaxial cyclic loading failure test, energy change during unloading process has different stress path from that during loading process. The mechanical properties of the two processes are different in nature.

The unloading confining pressure test of granite studies the impact of unloading confining pressure on rock strength and deformation characteristics with axial stress unchangeable. Test operators load a

certain stress (about 85% of peak strength, ensue that samples will not be damaged) in axial under confining pressure of 10MPa, 15MPa, 20MPa, 25 MPa and 30MPa respectively and lower confining pressure after making axial stress unchangeable using triaxial tester, until samples are damaged.
At confining pressure of 10MPa, 15 MPa, 20MPa, 25 MP and 30MPa, the axial pressure loads of samples are added to 180MPa, 200 MPa, 210MPa, 250MPa and 255 MPa successively and loading stops, so as to ensure the axial stress unchangeable. Meanwhile, the confining pressure is gradually unloaded. As shown in Figure7 (b), (c), (d), as confining pressure decreases evenly by 5MPa, the axial loads lower gradually and transverse stress enlarges gradually until rocks are damaged, showing that confining pressure improves the compressive strength of rock, effectively restricts the transverse crack and deformation as well improves the integrity of rock.

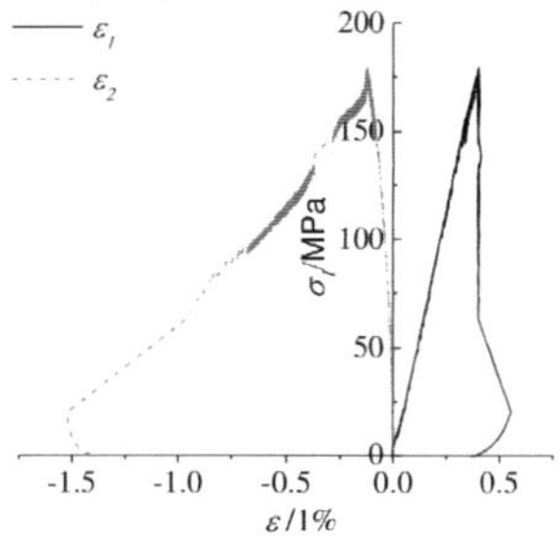

(a) Confining pressure 10MPa

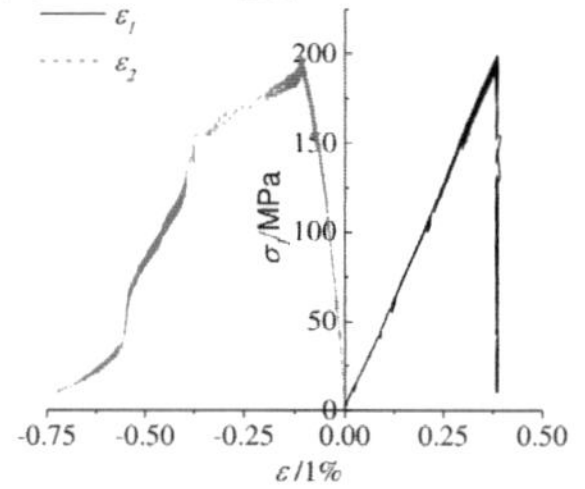

(b) Confining pressure 15MPa

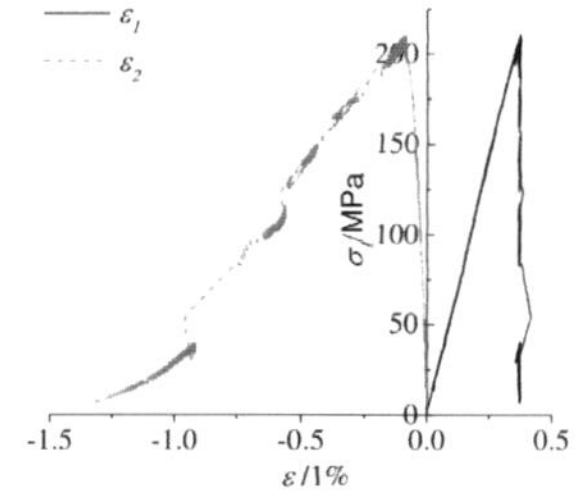

(c) Confining pressure 20MPa

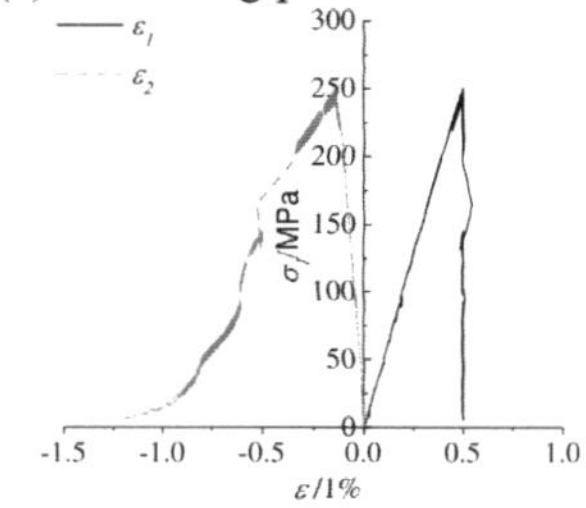

(d) Confining pressure 25MPa

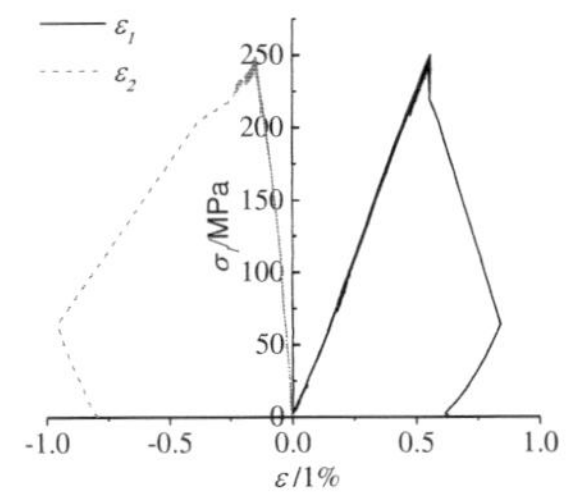

(e) Confining pressure 30MPa

Figure 7. Stress-strain curves of rock samples under axial unloading at different confining pressures

In the initial stage of unloading confining pressure, the transverse deformation of rock samples does not increase greatly and transverse stress presents a linear relation with confining pressure. The three change curves of transverse stress at confining pressure of 10MPa, 15MPa and 20MPa are basically in a line and two curves with confining pressure of 25MPa and 30MPa present similar deformation process, proving that initial deformation is under elastic deformation phase. With continuous decrease in confining pressure, transverse deformation increases to some extents and features plastic failure. Five deformation curves in Figure8 shows that the larger the initial value of confining pressure is, the larger the confining value during plastic failure will be. That is to say, samples failure occurs not during the process that confining pressure is totally removed but during unloading process.

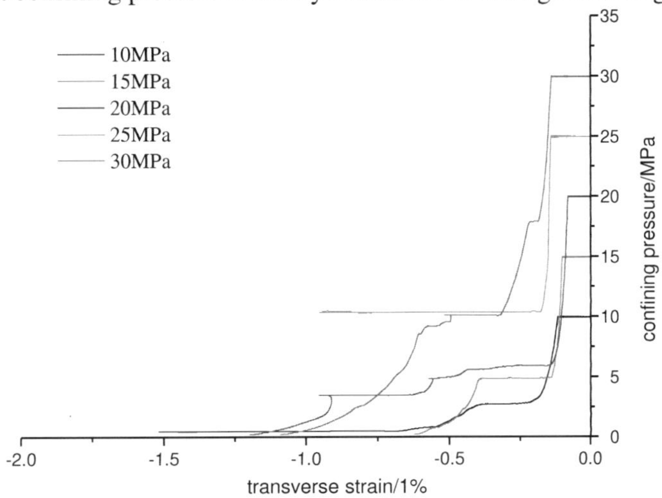

Figure 8. The confining pressure curves with transverse strain change of rock samples

Axial stress-confirming pressure change curve of granite is shown in Figure9. In the initial stage of unloading confining pressure, the axial stress descends with confining pressure decrease, yet the descending extent is relatively slow. After confining pressure descends for a while, axial bearing capacity of samples will gradually lower to the axial stress value loaded by instrument. Then samples are damaged and axial bearing capacity curve plunges rapidly. The process demonstrates that the axial bearing capacity dropped with a decreasing confining pressure in an essentially linear way, and the existence of confirming pressure played a role in stabilizing the axial bearing capacity. Also, optimization support is proven to be of vital importance to engineering project.

With axial displacement unchangeable, the energy to cause rock failure is elastic deformation energy stored inside rocks during the failure process of rock specimen under triaxial unloading process of confining pressure and there is no action from other external forces. Under triaxial stress state, when stress at a direction falls suddenly, rock failure will occur even thought the rock is under low stress state. However, due to low stress, energy value absorbed during rock failure is less. The process focuses on elastic energy release. If rock mass lacks effective support, energy released by protolith will be transferred into kinetic energy to crack rocks, leading to striking rock burst.

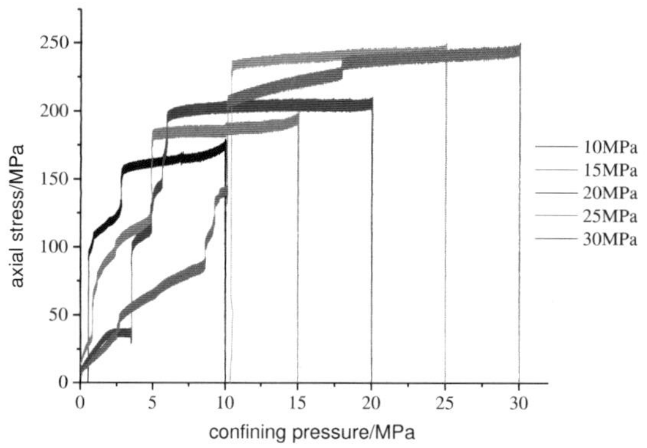

Figure 9. Variation curves of axial stress and confining pressure of granite

6. Conclusion

On the basis of research and analysis of energy evolution mechanism and confining pressure effect of granite under triaxial cyclic loading and unloading, the paper draws the following conclusions:

(1)As for the stress-strain curve of rock, the increase in confining pressure affects the change during pre-peak phase to some extents. In the ascending process of the confining pressure from 10MPa to 30MPa, the stress peaks increased by 50% and 22% respectively and the pre-peak weakening became more apparent; the area enclosed by the hysteresis loop corresponding to 30MPa diminished by nearly 60% than that corresponding to 10MPa, indicating a higher confining pressure prohibits rock mass from plastic deformation and shifts strain toward elastic deformation. In the post-peak phase, the larger the confining pressure is, the gentler the decrease curve will be.

(2)Pre-peak phase of rock focuses on energy accumulation. The proportion of elastic energy in input energy increases with the increase in loads. In the vicinity of the strength limit, the slope of dissipation energy increased to 1.6 from the original 0.7, rising by 128% or so. At the peak intensity, the elastic energy proportion reaches the peak and the dissipation energy proportion reaches the lowest point, showing that energy evolution begins to change before loads reach the peak intensity. Prior to failure, the dissipation energy grew at an accelerating rate, demonstrating stronger propagation and convergence of internal cracks.

In energy ratio curve, elastic energy takes up 0.7-0.9, much higher than that of dissipation energy, demonstrating that most mechanical energy from the outside are transferred into elastic energy and only 10%-30% of mechanical energy are transferred into other dissipation energy. Prior to stress peak, stress change impacts energy evolution process, revealing the distribution law of energy inside rock.

Under triaxial stress state, the axial bearing capacity dropped with a decreasing confining pressure in an essentially linear way, and the existence of confirming pressure played a role in stabilizing the axial bearing capacity. When the stress at a certain direction plunges, energy inside rock is released quickly and elastic energy transfers into kinetic energy to crack rocks, leading rock mass failure. Therefore, in actual engineering, it is essential to conduct effective pre-supporting measures on rock masses so as to get rid of potential safety risks.

7. Reference

[1] Xie Heping, Ju Yang and Li Liyun 2005 Criteria for Strength and Structural Failure of Rocks based on Energy Dissipation and Energy Release Principle *J. Chinese Journal of Rock Mechanics and Engineering.* **24** 3003-3010

[2] Xie Heping, Ju Yang and Li Liyun 2008 Energy Mechanism of Deformation and Failure of Rock Masses *J. Chinese Journal of Rock Mechanics and Engineering.* **9** 1729-1740

[3] Zhang Zhizhen and Gao Feng 2012 Research on Nonlinear Characteristics of Rock Energy Evolution under Uniaxial Compression. *J. Chinese Journal of Rock Mechanics and Engineering.* **6** 1198-1207

[4] Zhang Zhizhen and Gao Feng 2015 Experimental Study on Energy Evolution Characteristics of 3 Varieties of Rocks. *J. Journal of China University of Mining and Technology.* **3** 416-422

[5] Cai Meifeng, Ji Dong and Guo Qifeng 2013 Study of Rockburst Predication Based on In-site Stress Measurement and Theory of Energy Accumulation Caused by Mining Disturbance. *J. Chinese Journal of Rock Mechanics and Engineering.* **10** 1973-1980

[6] Zhang Zijian, Ji Hongguang and Zhang Yuezheng 2015 Study on Rockburst during Deep Mining Excavation in Linglong Gold Mine *J. Metal Mine.* **6** 134-138.

[7] Zhang Zhizhen and Gao Feng 2012 Experimental Research on Energy Evolution of Red Sandstone Samples under Uniaxial Compression. *J. Chinese Journal of Rock Mechanics and Engineering.* **5** 953-962

[8] Peng Ruidong, Ju Yang and Gao Feng 2014 Energy Analysis on failure of Coal under Cyclical Triaxial Loading and Unloading Conditions. *J. Journal of China Coal Society.* **2** 245-252

[9] Yang Shengqi, Xu Weiya and Su Chengdong 2007 Study on the Deformation Failure and Energy Properties of Marble Specimen under Triaxial Compression. **J. Engineering Mechanics. 1** 136-142

[10] You Mingqing and Hua Anzeng 2002 Energy Variation Analysis during failure Process of Rock Samples. *J. Chinese Journal of Rock Mechanics and Engineering.* **6** 778-781

[11] Fu Yufang, Qi Hongwei and Huang Mingli 2000 Numerical Tests on Confining Pressure Effect in Rock Failure Process. *J. Journal of Liaoning Technical University (Natural Science).* **5** 488-492

[12] Pan Xiuying and Liu Lianlian 2014 Research on Triaxial Compression Test of Granite after High Temperatures. *J. Urban Construction Theory Research: electronic edition.* **31** 378-289

[13] Zhao Zhonghu, Lu Rui and Zhang Guoqing 2006 Analysis on Energy Transformation for Rock in the Whole Process of Deformation and Fracture *J. Mining R& D.* **5** 8-11

[14] Zhang Zhizhen, Gao Feng, 2015. Confining Pressure Effect on Rock Energy. Chinese Journal of Rock Mechanics and Engineering. 6(1): 1-11.

[15] Dai Bing, Zhao Guoyan and Yang Chen 2016 Energy Evolution Law of Rocks in Process of Unloading Failure under Different Paths *J. Journal of Mining & Safety Engineering.* **2** 121-128

IOP Conf. Series: Materials Science and Engineering **362** (2018) 012019 doi:10.1088/1757-899X/362/1/012019

Removal of Free Fatty Acid from Plant Oil by the Adsorption Process

Tsair-Wang Chung, Yi-Ling Wu, and Shih-Hong Hsu

Department of Chemical Engineering and Center for High-Value Bioresource Industry Promotion, Chung Yuan Christian University, Taoyuan, Taiwan
E-mail: twchung@cycu.edu.tw

Abstract. The food oil refinery process for deacidification is ususally conducted by the neutralization after degumming. In this study, commercialized resins will be used as adsorbents to remove the free fatty acid (FFA) in food oil without using any solvent. Applying this environmental friendly green process, the energy efficiency will be increased and the waste water will be reduced compared to the traditional process. The selected adsorbent can be reused which may reduce the process cost. Instead of using alkali neutralization, the proposed process may reduce the concern of food oil security. The commercial resins A26OH and IRA900Cl were compared as adsorbents to remove the FFA in deacidification for refinery of food oil without adding any alkali chemicals. This process will be conducted to remove the FFA form peanut oil in this study. Besides, this study will get the adsorption isotherms for one of the better sorbents of A26OH or IRA900Cl to remove FFA from peanut oil under 25, 35, and 45°C. The Langmuir and Freundlich isotherm models were compared to fit the experimental data. The obtained isotherm data is important for the adsorption system design.

1. Introduction

Human may obtain the nutrients such as carbohydrates, proteins and lipids by food. Lipids are the main source of metabolic energy for human body. The main sources of edible lipids are the oils and fats, including plant and animal fats. In addition to glycerides, there are impurities such as phospholipids, free fatty acids (FFA) and others in crude oil that causes deterioration of the quality and curtailment of shelf life of oil whether plant oil or animal oil. Therefore, these impurities must be removed from crude oil through refining process. [1-3] There are quite a few refining process in business despite commercially available. In this study, using commercial resin as adsorbent in the absence of adding solvent for the adsorption process to remove FFA in plant oil. The application of environmentally friendly green process also enhances the use of energy, reduces the use of toxic substances and the generation of waste while improving efficiency and product quality, and emphasizes recyclable and reusable designs to mitigate environmental pollution load and impact, but also reduce the concern about food safety. [4-6]

2. Deacidification by ion exchange resins

The refining process of edible oil is conducted by steps of degumming, deacidification, bleaching and deodorization. The presences of FFA in edible oil raise the possibility of rancidity, even small amounts of fatty acids that reduce the quality of the oils. [7] Moreover, the presence of free fat causes generation of corrupt odours and accelerates oxidation process. [8] In recent years, many studies use strong anion exchange resins to adsorb oleic acid or FFA in plant oil, but all of the studies were operated continuously in packed beds with alcoholic solvents.

Hadiah used a strong anion exchange resin, IRA900Cl, for adsorption removal of free fatty acids in Kapok seed oil, Bagilumbang seed oil and Candle nut oil under a packed bed shown in Figure 1 with n-hexane as solvent under room temperature to obtain the impact of the effect of different plant oil variety on removal of acid value. The results are shown in Table 1, the removal rate of acid value is 95%, and the effect of deacidification is very good. And also recovery the ion exchange resin IRA900 Cl after desorption and then carry out the deacidification of Bagilumbang seed oil for repeated 5 times. Repeated uses of 3 times of IRA900Cl within the deacidification effect are maintained at a fine level, the results are shown in Table 2. Additionally found that the use of IRA900 Cl for deacidification can achieve bleaching effect. [9]

Cren used a strong anion exchange resin, A26OH, to remove oleic acid in an oleic acid ethanoic solution under packed bed operation at room temperature to investigate the effect of flow rates and initial concentration of oleic acid. The results shown in Table 3, the removal rate of acid value is 97%, and the effect of deacidification is very well. And also found that the higher initial concentration of oleic acid or the lower flow rate, the higher removal rate of oleic acid. Conversely, the higher initial concentration of oleic acid or the higher flow rate, the higher adsorption amount of oleic acid [10]

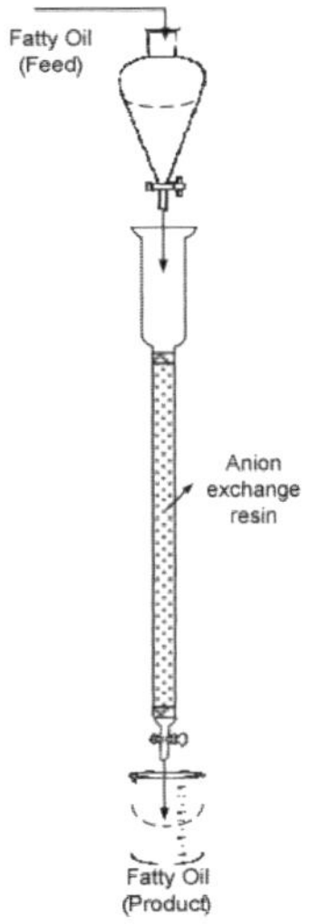

Figure 1. Schematic of packed bed set up for adsorption deacidification of different plant oils by using ion exchange resin IRA900Cl. [9]

Table 1. The deacidification results of adsorption different plant oils by using ion exchange resin IRA900Cl. [9]

Oil variety	Initial acid value of oil (mg KOH/g)	Acid value of oil after adsorption (mg KOH/g)	Removal rate of acid value (%)
Kapok seed oil	18.12	0.59	96.74
Bagilumbang seed oil	13.20	0.56	95.76
Candle nut oil	10.80	0.54	95.00

Table 2. The deacidification results of by recovery ion exchange resin IRA900Cl after desorption and then carry out adsorption of Bagilumbang seed oil for repeated 5 times. [9]

The number of reuses	Acid value of oil after adsorption (mg KOH/g)	Removal rate of acid value (%)
1	0.56	95.76
2	0.59	95.53
3	0.59	95.53
4	5.60	57.58
5	10.32	21.82

Table 3. The deacidification results of adsorption different plant oils by using ion exchange resin IRA900Cl. [10]

Flow rate (mL/min)	Initial concentration of oleic oil (wt%)	Operation time (min)	Removal rate of acid value (%)	Adsorption amount (g acid/g dry resin)
19.4	4.21	198	97.7	1.23
19.5	6.13	155	98.3	1.29
30.0	4.22	136	97.8	1.17
30.0	6.34	95	98.2	1.22
16.7	4.90	180	98.1	1.18
33.2	4.90	120	97.9	1.13
26.3	3.59	176	97.7	1.13
24.5	6.50	124	98.2	1.22
24.9	5.19	135	98.5	1.23
24.0	5.26	133	97.8	1.24
24.3	5.18	132	97.9	1.18

3. Experimental method

3.1. Materials
Commercial ion exchange resin IRA900Cl (Amberlite®IRA900 chloride form) and A26OH (Amberlyst®A26 hydroxide form), and oleic acid with format of 65.0-88.0%, Ph. Eur. Grade were purchases from Sigma-Aldrich. Commercial oil Special Grad Peanut Oil was purchase from a manufacturer in Taoyuan Taiwan. Fresh Squeezed under Low-Temperature Peanut Oil was purchase from Master Chen Oil Workshop in Changhua Taiwan.

3.2. Pretreatment of oil
In order to increase the initial acid value of oil, add oleic acid directly into peanut oil. Since the content of free fatty acids in oils is often defined as the content of oleic acid in oils, the method of adding oleic acid directly into oils is also widely used in many studies.[9-11] In order to reduce the initial acid value of the oil, there are some manufacturers add good quality oil with low acid value into high-acidity oil products with poor quality to reduce the cost and acid value of their oil products. In this study, the peanut oil product from Master Chen Oil Workshop which has been through membrane filtration was added into commercial peanut oil that had not been filtered.

3.3. Pretreatment of adsorbents
Put 3 grams of ion exchange resin on a Petri dish covered with aluminium foil paper, placed into a constant temperature oven at 50°C for 3 hours to remove the moisture on the surface of adsorbent.

3.4. Adsorption equilibrium time
Loading commercial peanut oil, which acid value was unadjusted into 20 mL sample bottle, and add 1 wt% of ion exchange resin IRA900Cl and A26OH respectively. Then placed in a constant temperature circulating water bath under 35°C, the stirring speed was set to 550 rpm to operate 0.5, 1, 1.5, 2, 2.5, 3 hours adsorption respectively. After the adsorption, the ion exchange resins are separated from the oil by centrifugation. Finally, detecting the acid value of the deacidified peanut oil from which the adsorbent has been removed and calculate the adsorption efficiency index to compare the effect of the adsorbent. The adsorption equilibrium time obtained for this test is taken as the adsorption time of all subsequent experiments in this study.

3.5. Isothermal adsorption curve

Loading commercial peanut oil, which acid value was adjusted to 1, 3, 5, 8, 10, 13 mg KOH/g into 20 mL sample bottle respectively, and add 1 wt% of ion exchange resin A26OH. Then placed in a constant temperature circulating water bath under 35°C, the stirring speed was set to 550 rpm. The adsorption time was set as the adsorption equilibrium time obtained from *Section 3.4*. After the adsorption, the ion exchange resins are separated from the oil by centrifugation. Finally, detecting the acid value of the deacidified peanut oil from which the adsorbent has been removed and calculate the amount of adsorption; draw it into isothermal adsorption curve and further fit Langmuir and Freundlich adsorption isotherm models to understand the adsorption patterns.

4. Results and Discussion

4.1. Pretreatment of adsorbents

The specific surface areas of adsorbents with and without the pretreatment were measured by Laser Diffraction Method and the results are listed in Table 4. It can be seen from the table that the specific surface area of IRA900Cl is larger than that of A26OH, while the specific surface area of pretreated IRA900Cl is smaller than that of the one without pretreatment. However, the specific surface area of A26OH is not significantly changed due to the pretreatment. The surface picture results of the adsorbents with and without pretreatment by transmission electron microscope are shown in Figure 2. It can be seen from the figure that the treated adsorbent is cracked or even broken, and the surface of the adsorbent is smooth.

4.2. Adsorption equilibrium time

1 wt% ion exchange resin A26OH and IRA900Cl were used to adsorption FFA from pretreated peanut oil respectively under 35°C. Results of equilibrium concentration affected by adsorption time are shown in Figure 3. Results of removal rate of FFA affected by operation time are shown in Figure 4.

As shown in Figure 3. After 0.5 hours of adsorption, the equilibrium concentration decreased from 15.75 mg/g to 13.55 mg/g; after 1 hour of adsorption, the equilibrium concentration decreased slightly to 12.45 mg/g and maintained for 1.5 hours; after 2 hours, the decrease of equilibrium concentration was less. After the equilibrium concentration dropping to 11.7 mg/g and few changes to 3 hours, the adsorption equilibrium time of A26OH to could be regarded as 3 hours. The IRA900Cl adsorption for 0.5 hour equilibrium concentration from 15.75 mg/g down to 15 mg/g, is a relatively large drop; after 1 hour of adsorption, the equilibrium concentration decreased only to 14.8 mg/g; after two hours of adsorption, the equilibrium concentration also dropped to only 14.55 mg/g and maintained to 3 hours, so that IRA900Cl adsorption equilibrium time can be considered as 3 hours. A26OH was found to up to 25.5% of FFA removed from Figure 4, removal rate of FFA by IRA900Cl was less than 10%, A26OH adsorption was significantly higher than that of IRA900 Cl, so the future use of A26OH as the experimental adsorbent in this study.

4.3. Isothermal adsorption curve

The ion exchange resin A26OH was used to adsorb peanut oil with different initial concentrations of FFA under 25, 35 and 45°C. The isothermal adsorption curve results were fitted to the Langmuir Isothermal adsorption model and the Freundlich isothermal adsorption model shown in Figure 5 and the fitting contanst results were given in Table 5. It was found from Table 5 that coefficient of determination (R^2) for fitting Langmuir isothermal adsorption model were higher than Freundlich model under all the operation temperature; accordingly, Langmuir isotherm adsorption model was suitable to describe the adsorption for ion exchange A26OH to adsorb FFA from peanut oil. The maximum adsorption capacities (Q_0) were found in Table 5, whether at any operation temperature, value is the same of 555.56 mg/g. And Freundlich experience index (n) was found that n> 1; hence the adsorption for ion exchange A26OH to adsorb FFA from peanut oil was favorable adsorption.

IOP Conf. Series: Materials Science and Engineering **362** (2018) 012019 doi:10.1088/1757-899X/362/1/012019

Table 4. The specific surface areas results of adsorbents with and without the pretreatment by Laser Diffraction Method

Adsorbents	Untreated IRA900Cl	Treated IRA900Cl	Untreated A26OH	Treated A26OH
Specific surface areas (m^2/g)	673.20	608.46	492.67	494.66

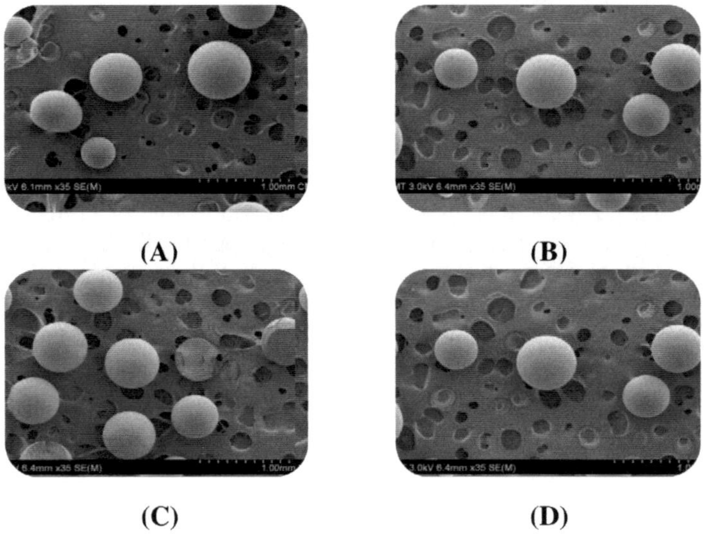

Figure 2. The surface picture results taken by transmission electron microscope (SEM) of (a) untreated IRA900Cl; (B) treated IRA900Cl; (C) untreated A26OH and (D) treated A26OH.

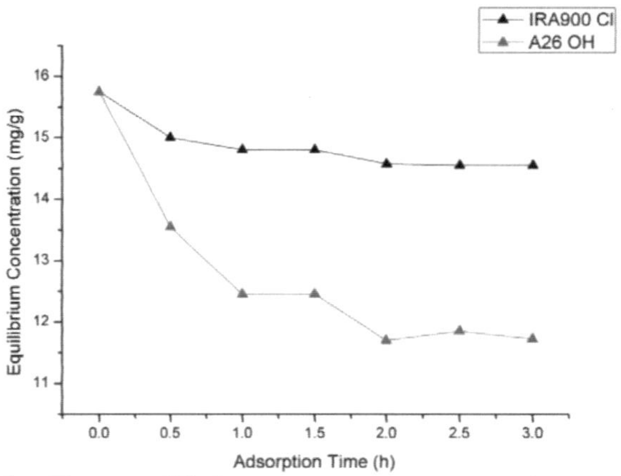

Figure 3. The equilibrium concentration results at different operation time with IRA900Cl and A26OH as adsorbents to adsorption FFA in peanut oil.

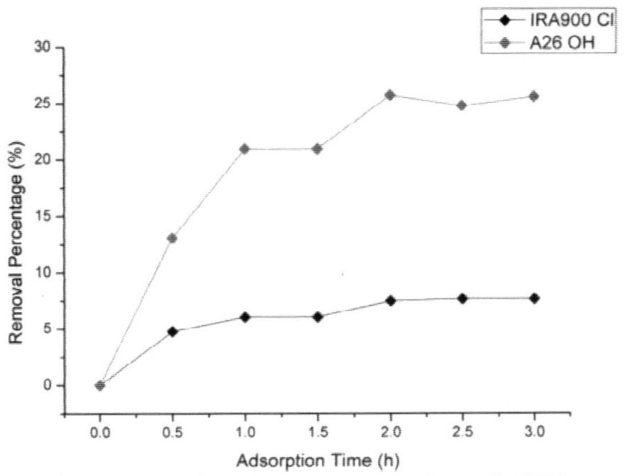

Figure 4. The removal percentage results of FFA at different operation time with IRA900Cl and A26OH as adsorbents to adsorption FFA in peanut oil.

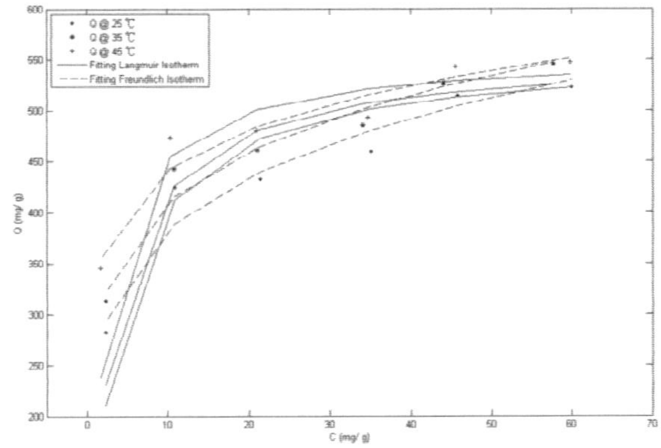

Figure 5. The isothermal adsorption curve and fitted curve results of Langmuir and Freundlich isotherm models by A26OH adsorbed peanut oil at 25, 35 and 45°C, respectively

Table 5. The regression contansts of Langmuir and Freundlich isotherm models by A26OH adsorbed peanut oil at 25, 35 and 45°C, respectively

Operation Temperature (°C)	Fitting Langmuir isotherm model			Fitting Freundlich isotherm model		
	Q_0	K_L	R^2	n	K_F	R^2
25	555.56	0.26	0.99	5.52	252.35	0.95
35	555.56	0.31	0.99	6.04	279.64	0.97
45	555.56	0.44	0.99	8.03	331.44	0.95

5. Conclusion

In this study, commercial ion exchange resins A26OH and IRA900Cl were used as adsorbents in the deacidification step of plant oil refining process, and the free fatty acids in commercial peanut oil were adsorbed and removed without adding any solvent. By comparing the adsorption performance of different ion exchange resins, it was found that addition of 1 wt% adsorbent under 35°C resulted in removal of up to 25.5% free fatty acids by A26OH and less than 10% removal of free fatty acids by IRA900Cl. Therefore, adsorbent performance by A26OH is better than by IRA900Cl. The present study also described the adsorption of free fatty acids in the peanut oil using A26OH ion exchange resin. The isothermal adsorption curves at 25, 35 and 45°C respectively shown that adsorption at all temperatures is more consistent with Langmuir isothermal model, and the adsorption also found that were favourable adsorption at all temperatures. For future study, the dynamic adsorption studies will be conducted in a packed-bed column and the data will be useful to design the real application column for oil refinery.

6. References

[1] Dijkstra A J 2010 European Journal of Lipid Science and Technology 112 pp1178-1189

[2] Mei L, Wang L, Li Q, Yu J and Xu X 2013 *Journal of the Science of Food and Agriculture* 93 pp2820-2828

[3] De B K and Patel J D 2010 *Journal of Oleic Science* 59 pp121-125

[4] Bhosle B M and Subramanian R 2005 *Journal of Food Engineering* 69 pp 481-494

[5] Be B K and Patel J D 2010 *Journal of Oleic Science* 59 pp121-125

[6] García A, Álvarez S, Riera F, Álvarez R and Coca J 2006 *Journal of Food Engineering* 74 pp516-522

[7] Vaisali C, Charanyaa S, Belur P D and Regupathi I 2015 International Journal of Food Science & *Technology* 50 pp 13-23

[8] Chaiyasit W, Elias R J, McClements D J and Decker E A 2007 Critical Reviews in Food *Science and Nutrition* 47 pp 299-317

[9] Hadiah F, Irawati, Prakoso T, Subagjo and Soerawidjaja T H 2014 Proceedings of the 5th Sriwijaya International Seminar on Energy and Environmental Science & Technology Palembang pp 71-74

[10] Cren E C, Filho L C, Silva E A and Meirelles A J A 2009 *Separation and Purification Technology* 69 pp 1-6

[11] Ilgen O 2014 *Fuel Processing Technology* 118 pp 69-74

IOP Conf. Series: Materials Science and Engineering **362** (2018) 012020 doi:10.1088/1757-899X/362/1/012020

A New Thermal Treatment Process of Low Value Volcanic Glass towards the Production of Expanded Material and its Use on CNTs' Synthesis as Substrate Material

Panagiotis M. Angelopoulos [1*], Michail Samouhos[1], Maria Taxiarchou[1], P. Tsakiridis[1], John Haggman[2], Paul Joyce[2]

[1]School of Mining and Metallurgical Engineering, Laboratory of Metallurgy, National Technical University of Athens, Greece 9, Iroon Polytechniou Street, 157 80 Zografou, Athens, Greece

[2]Perlco Pty Ltd, Suite 701, Level 7, 121 Walker str, North Sydney NSW 2060, Australia

*pangelopoulos@metal.ntua.gr

Abstract. Pitchstone is a naturally occurring volcanic glass that contains considerable amount of chemically bound water (> 6 % wt). Due to its high water content, its direct thermal processing in conventional expansion furnaces towards the production of lightweight material, similar to expanded perlite, is practically impossible. In the current research paper a sophisticated 2 stage process is presented that consists of a partial dehydration and an expansion stage towards the production of high quality expanded material. After proper treatment, low-value volcanic glass is transformed to frothy, lightweight material of closed external surface and apparent density of 52 $kg \cdot m^{-3}$ that can be used in various branches of the industry. The material produced is used as substrate for the development of multiwall CNTs through CVD method. Dense multiwall CNT clusters were identified on expanded pitchstone surface, thus rendering the material suitable for such application.

1. Introduction

Carbon nanotubes (CNTs) have attracted great research interest since they uniquely combine important properties, including chemical stability, high surface area and extraordinary mechanical strength and electric conductivity, as a result of their cylindrical nano-architecture. CNTs are currently used in a wide range of applications including electronic and optoelectronic devices, energy storage, high strength composites, sensors and catalysts [1].Chemical vapour deposition method (CVD) constitutes a widely applied technique for CNTs synthesis due to its low cost, low temperature and relative equipment simplicity, in comparison to other existing methods i.e. carbon-arc discharge and laser ablation of carbon. Hydrocarbon (carbon source), catalyst and catalyst support (growing media) constitute key materials that are involved in CNTs production through CVD method. Commonly used CNT precursors are purified carbon products, namely methane, benzene and acetylene, while concerning substrates graphite, quartz silicon, silicon carbide, silica, alumina, alumina-silicate have successfully being used. Fe, Co and Ni metals are used as catalysts since they provide with high solubility and diffusivity of carbon atoms at elevated temperatures [2]. An alternative natural CNTs' precursor, that successfully have been used for synthesis of single and multi-walled CNTs constitute camphor ($C_{10}H_{16}O$) that is obtained from cinnamomum camphora tree. Unlike petroleum by-products, camphor is eco-friendly and can be easily cultivated in as much quantity as required [3]. Recently, expanded perlite proved to be an effective substrate for multiwall CNT production with remarkable production yield [4]. Pitchstone is a naturally occurring volcanic glass, similar to perlite and obsidian. The major difference between these materials is their water content; obsidian, perlite and pitchstone water content range between 0-2, 2-5 and >6 wt. %, respectively. Due to the presence of chemically bound water, these materials expand when heated at a temperature that exceeds their softening point as contained water vaporizes and pushes its way out of the softened grains, resulting in the formation of granular, cellular material of significant low density. Since expansion phenomenon is affected by chemical composition, water content and also crystalline content of the glass, expansion potential is different for obsidian, perlite and pitchstone. Perlite constitutes best performance material, and after proper single step heating in expansion furnace at temperature above 800 °C attains loose bulk density (LBD) as low as 30-150 $kg \cdot m^{-3}$ [5]. In its expanded form, perlite become white, frothy and cellular, with a wide network of air bubbles in its structure. Expanded perlite is lightweight, inorganic, chemically inert, fire resistant, with superior thermal and sound insulating properties, while its properties are invariant in time, and is widely used in construction, horticulture, filtration and other applications.

High water content of pitchstone constitutes a barrier towards its transformation to lightweight material in traditional gas-fired expansion furnaces. As reported in a recently published work, in order to transform it to high-quality expanded material, sophisticated thermal treatment has to be implemented which combines partial dehydration in laboratory oven and expansion in vertical electric furnace [6]. One of the biggest pitchstone occurrence is located in north of Chillagoe, in north Queensland, Australia. The occurrence is owned by Perlco Pty Ltd and may contains 700 Mt resource [7]. 100 Mt of pitchstone exists, and on annual basis, 30% of pitchstone is crushed into fine stone-dust with particle size less than 500 μm [8]. As a consequence of its fineness and its relatively high water content, at present, this stone-dust is stockpiled since it has no economic value [9]. However, recent researches have proved the pozzolanic activity of pitchstone fines thus rendering suitable their use as cement replacement in concrete [9]-[11].

The objectives of the current research article are dual; (a) a processing method for transformation of low value fine pitchstone to high quality expanded material is presented through a two steps thermal treatment method for effective partial dehydration and expansion of pitchstone grains, and (b) the expanded material was used as substrate for the development of multiwall CNTs. Successful application of the new material will provide with a new application of low value volcanic glasses.

2. Materials and methods

2.1. Pitchstone characterization

The chemical analysis of raw pitchstone was performed using Energy Dispersive X-Ray Fluorescence instrument and the Al_2O_3 content was defined by flame Atomic Absorption Spectroscopy after dissolution in a $HF/HClO_4/HNO_3/HCl$ mixture. Pitchstone sample mainly consists of silicon and aluminum oxide (70.7 and 13.1 %, respectively), while its content in K_2O and Na_2O is 3.5 % wt. each. Minor contents in Fe_2O_3 and CaO were identified in amount 0.96 and 0.92 % wt, respectively. A commonly applied classification method of volcanic matter constitutes TAS classification. TAS classification can be used to assign names to many common types of volcanic rocks based upon the relationship between the combined alkali content and the silica content [12]. According to the chemical composition of the pitchstone sample, it is classified as rhyolite.

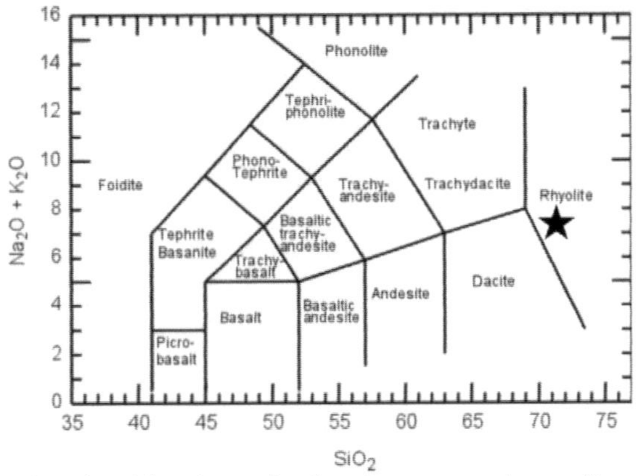

Figure 1 TAS diagram for the classification of volcanic matter. According alkalis and silica content of pitchstone sample it is classified as rhyolite (star sign)

Loss on Ignition (LOI) constitute weight loss after heating at 950 °C for 4 h as suggested by Perlite Institute, and is attributed to release of different forms of water. LOI of pitchstone sample was found to be 7.41 % wt. The mineralogical composition was examined on X-Ray diffraction device with Cu-Kα (Ni filtered) radiation (2 theta range from 5 to 75° and 0.02°·sec^{-1} step), and indicated the dominance of amorphous phase with minor quartz and feldspar content. The rough XRD pattern of pitchstone sample is characteristic for non-crystalline materials, while the hump between 15 and 30 2 theta (degree) is characteristic for volcanic glasses.

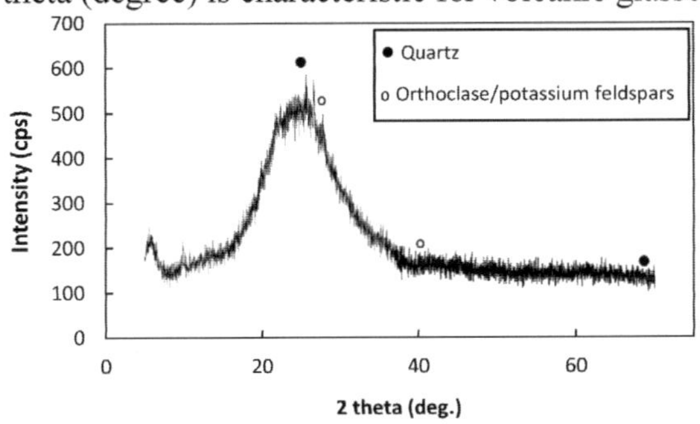

Figure 2 XRD pattern of pitchstone sample

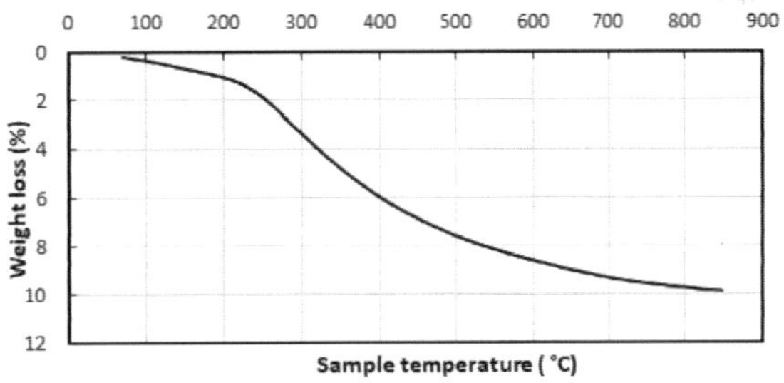

Figure 3 Weight loss of pitchstone in terms of heating temperature (thermogravimetry)

Thermogravimetric analysis was performed at a Labsys DTA/DSC Setaram device at sample heating rate 10 °C per min and under atmospheric conditions, and revealed two stages of weight loss, which are attributed to moisture and loosely held (up to 200 °C), and chemically bound water (over 200 °C) release.

2.2. Pitchstone processing towards production of expanded material

Raw pitchstone has been subjected to mechanical and thermal treatment towards the production of lightweight expanded material. Raw sample has been crushed, milled and sieved in order to produce a grade with particle size range between 100 and 300 μm. Equipment employed for this purpose consists of jaw crusher, rotor beater mill and continuous sieve device with aperture size of 100 and 300 μm. Then, sample heat treatment took place at 3 different units. First, moisture removal took place through sample heating at 105 °C overnight (16 h). Scoping the partial dehydration of the sample, pitchstone has been heated at 300 °C for 1 h. The LOI of partially dehydrated sample reached the desirable value of 3.92%. Finally, the partially dehydrated sample has been fed in the vertical electric furnace. The furnace was set to 1100 °C and the particle heating time did not exceed 4 s. It is noteworthy that the use of vertical electric furnace is inevitable since traditional gas fired furnaces are incapable to sufficiently treat such raw material due to its fineness [5].

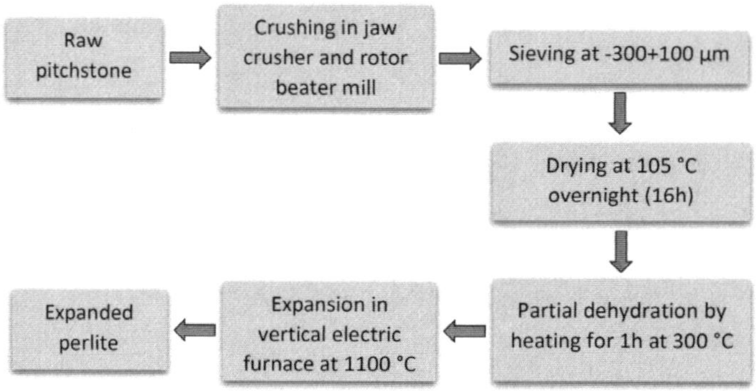

Figure 4 Mechanical and thermal treatment route of pitchstone towards the production of expanded material

2.3. CNTs' synthesis method

For multiwall CNTs' synthesis, a commonly applied synthesis process, namely, chemical vapour deposition (CVD) method has been applied. Camphor was used as carbon source, ferrocene as catalyst and expanded pitchstone as substrate. 100 gr of camphor and 5 gr of ferrocene were placed at the narrow part of quartz tube (diam=5 cm), which was heated at 200 °C. The vapors produced where drifted by mild nitrogen gas stream to the wide quartz tube (diam=10 cm) where expanded pitchstone (800 cm^3, 40 gr) was spread and heated peripherally at 800 °C by heating resistances. The synthesis duration was 30 minutes. After growth of CNTs on expanded pitchstone the product was diluted in 4N NaOH solution for 8 h, in order to dissolve the substrate and purify the CNTs.

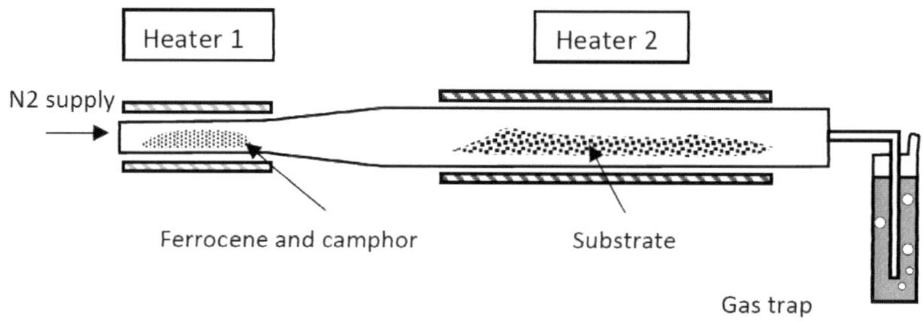

Figure 5 Experimental setup for CNTs' synthesis through CVD method

3. Results and discussion

3.1. Properties of expanded pitchstone

The sophisticated mechanical and thermal processing of low value pitchstone resulted in the production of high quality expanded material. The grains produced are cellular, white in colour and lightweight with LBD of 52 $kg \cdot m^{-3}$. The angular and non-symmetric raw pitchstone grains were transformed to expanded material of high sphericity, closed surface porosity and cellular structure, as shown in SEM images of Figure 6. Expanded material is characterised by high expansion ratio and low fragments content. It is worth noting that expansion experiments were also performed to non-dehydrated material, and the product obtained LBD value equal to 350 $kg \cdot m^{-3}$, thus rendering partial dehydration of pitchstone beneficial towards the production of high quality lightweight material.

IOP Conf. Series: Materials Science and Engineering **362** (2018) 012020 doi:10.1088/1757-899X/362/1/012020

Figure 6. SEM micrograph of raw (a), and expanded pitchstone (b, c)

3.2. Properties of produced CNTs

The morphology of synthesized CNTs was investigated using scanning electron (SEM) and high resolution transmission electron microscopy (HR-TEM). As depicted in Figure 7a, a pitchstone grain has been covered by the synthesized CNTs. CNTs are well dispersed over the grain surface while in certain locations the growth is well-developed in the vertical direction. After CNTs purification through dilution in NaOH solution, the identified CNTs were found to be orthogonal and carpet type that macroscopically compose clusters and tangles (Figure 7b-d). As can be seen in Figures 7b and c, the dissolution of pitchstone matrix was not complete since pitchstone fragment are still present. It is thus proposed the increase of NaOH concentration and in parallel the solution heating for complete dissolution of pitchstone towards the production of pure CNTs. In Figure 7d, a CNTs' cluster is observed that encloses catalyst particles and tiny pitchstone fragments. In HR-TEM Figure 7e the characteristic multilayer morphology of multiwall CNTs is identified, while in Figure 7f a detailed view of CNTs produced is given, with the dark color denoting the catalyst particles

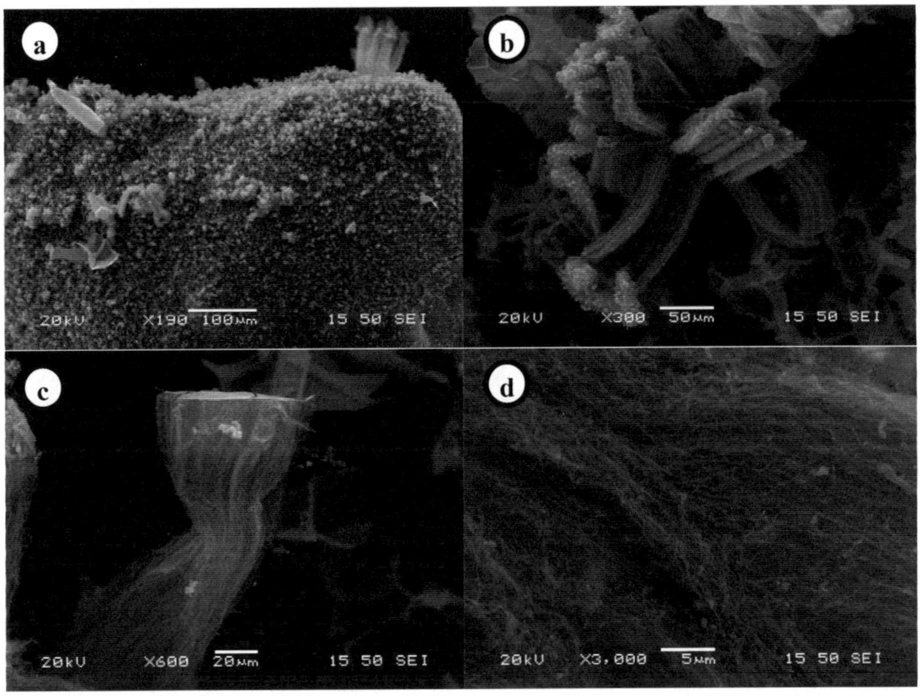

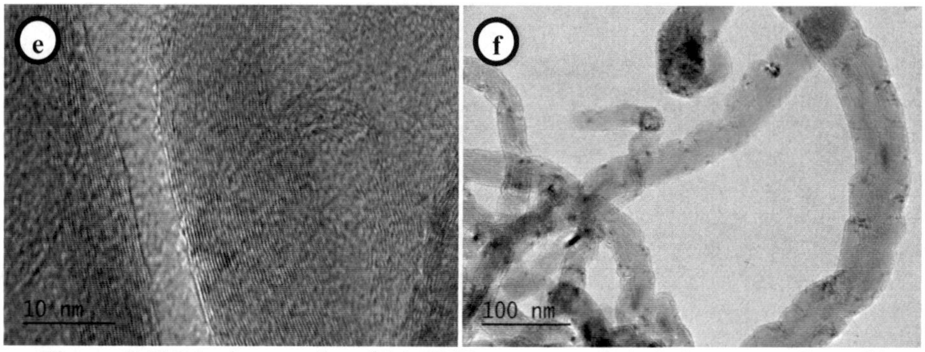

Figure 7 SEM micrographs of (a) CNTs on expanded pitchstone surface, (b) orthogonal growth CNTs, (c-d) CNTs clusters, and TEM micrographs of produced CNTs (e-f)

4. Conclusions

In the current research article a sophisticated thermal treatment method of low value volcanic glass towards their transformation to high quality expanded material is presented. The method incorporates primary sample heating at relative low temperature (300 °C) in order to reduce its water content below 4 wt. % , and subsequently expansion of the sample in a vertical electric furnace at 1100 °C. The material produced is frothy, granular with closed surface porosity and LBD value of 52 $kg \cdot m^{-3}$. Furthermore, expanded material produced was used as substrate for the development of multiwall CNTs through CVD method using camphor as precursor. CNTs were purified from substrate through dissolution of pitchstone by NaOH concentrate. It is shown that CNTs were developed at the majority of expanded pitchstone surface, while the CNTs structures vary due to surface inhomogeneity and thus catalyst polydispersity at substrate surface. It is thus shown here, that the expanded material produced though sophisticated treatment of low value volcanic glass can be used for high-end nanotechnology application such as substrate for CNTs synthesis.

5. Reference

[1] Hou P, Du J, Liu C, Ren W, Kauppinen E and Chend H-M 2017 *Applications of carbon nanotubes and graphene produced by chemical vapor deposition. MRS Bullet.* **42** (11), 825-833

[2] Kumar M and Ando Y 2010 *Chemical vapor deposition of carbon nanotubes: a review on growth mechanism and mass production*. J. Nanosci. Nanotechnol. **10** (6):3739-58.

[3] Kumar M, Zhao X, Ando Y, Lijima S, Sharon M and Hirahara K 2002 *Carbon nanotubes from camphor by catalytic CVD*, Mol. Cryst. Liq. Cryst. **387** (1), 117- 121

[4] Pilatos G, Samouhos M, Angelopoulos P, Taxiarchou M, Veziri C, Hutcheon R, Tsakiridis P and Kontos A 2016 *Carbon nanotubes growth on expanded perlite particle via CVD method: The influence of the substrate morphology*, Chem. Eng. Sci. **291**:106-114.

[5] Angelopoulos P, Gerogiorgis D.I and Paspaliaris I 2013 *Model-Based Sensitivity Analysis and Experimental Investigation of Perlite Grain Expansion in a Vertical Electrical Furnace*, Ind. Eng. Chem. Res. **52**(50), 17953–17975

[6] Angelopoulos P, Taxiarchou M, Paspaliaris I, Haggman J and Joyce P *2017 Implementation of indirect heating technology towards the production of expanded microspheres from low value pitchstone fines.* Proceeding of the Proc. of The IRES International Conf, Bucharest, Romania, 24th-25th October 2017, 5-10

[7] Breese R, and Barker J, "*Perlite; Industrial Minerals and Rocks*", 6th Edition, Society for Mining, Metallurgy, and Exploration, pp. 735-74, 1994.

[8] Tuladhar R, Sexton A, and Joyce P, *"Durability of concrete with pitchstone fines as a partial cement replacement",* Proc. of Concrete Institute of Australia's Biennial National Conference 16-18 October 2013, Gold Coast, QLD, Australia.

[9] Ray A, Sriravindrarajah R, Guerbois J, Thomas P, Border S, Ray H, Haggman J and Joyce P 2007 "*Evaluation of Waste Perlite Fines in the Production of Construction Materials*." J Therm. Anal. Calorim. **88** (1) pp. 279-283

[10] R. Tuladhar, M. Smith, G. Pandey, P. Joyce, "*Use of pitchstone fine as a partial replacement of portland cement for sustainable concrete*", Proc. of 2011 International Concrete Sustainability Conference, 2011, Boston, USA.

[11] K. Vessalas, A. Ray, P. Thomas, R. Sri Ravindrarajah, P. Joyce, and J. Haggman 2008 "*Pitchstone Fines Pozzolanic Activity Assessment as Partial Portland Cement (PC) Replacements*." Aust. Ceram. Soc. **44** (1) pp 7-12

[12] Le Maitre RW, Streckeisen A, Zanettin B, Le Bas MJ, Bonin B, P. Bateman P, Bellieni G, Dudek A, Efremova S, Keller J, Lamere J, Sabine PA, Schmid R, Sorensen H, and Woolley A, Igneous Rocks: A Classification and Glossary of Terms, Recommendations of the International Union of Geological Sciences, Subcommission of the Systematics of Igneous Rocks. Cambridge University Press, 2002. ISBN 0-521-66215-X

Acknowledgments

Panagiotis M. Angelopoulos would like to acknowledge the financial support of IKY. The project was supported by the program "IKY Fellowship of excellence for postgraduate studies in Greece-Siemens"

IOP Conf. Series: Materials Science and Engineering **362** (2018) 012021 doi:10.1088/1757-899X/362/1/012021

Preparation of $ErMnO_3$ by Sol-gel Method and its Photocatalytic Activity for Removal of Methyl Orange from Water

X Y Xie, J N Yang, L L Yu, J Y Min, D D Sun, P S Tang and H F Chen

Department of Materials Engineering, Huzhou University, Huzhou 313000, China
E-mail: tangps@163.com, headder@zjhu.edu.cn

Abstract. The single phase perovskite $ErMnO_3$ was synthesized using $Er(NO_3)_3$, manganese acetate, citric acid and urea by a facile sol-gel method. The gel of $ErMnO_3$ precursor was kept for 36 hours in 100 °C oven to get the xerogel. Then, the xerogel was calcined at 800 °C for 12 hours in muffle furnace to prepare single phase $ErMnO_3$. The prepared sample was characterized by thermogravimetry differential scanning calorimetry (TG-DSC), X-ray diffraction (XRD), scanning electron microscopy (SEM) and Fourier transform infrared spectroscopy (FT-IR). Under ultraviolet light, the photocatalytic activity of $ErMnO_3$ was studied with methyl orange of 20 mg/L as the simulated sewage. The results show that the $ErMnO_3$ sample particle size distribution is relatively uniform, the average grain size is mainly around 100 nm. The photocatalytic experiment demonstrates that $ErMnO_3$ is highly photocatalytic activity for removal of methyl orange from water. When methyl orange of 20 mg/L is degraded for 120 min in the presence of $ErMnO_3$, the degradation rate of methyl orange can reach about 95%. The degradation of methyl orange accords with first order kinetic model in presence $ErMnO_3$ sample, and the apparent rate constant is 0.022 min^{-1}.

1. Introduction

As a new type of semiconductor material, the application of perovskite rare-earth manganate $ErMnO_3$ in photocatalytic degradation of sewage has not been reported [1-2]. At present, the methods of prepared $ErMnO_3$ are mainly hydrothermal method [3] and solid state reaction method [4-5] and so on. The sol-gel has the advantages which chemical reaction is easily controlled, with high purity, small granularity and good uniformity of powders [6-8]. In addition, the preparation of perovskite $ErMnO_3$ by sol-gel method has never been reported. Therefore, it is of great significance to prepare $ErMnO_3$ by sol-gel method and to investigate its photocatalytic activity for removal of methyl orange from sewage [9-10].

In this paper, using erbium nitrate, manganese acetate, citric acid and urea as the main raw materials, $ErMnO_3$ was prepared by a facile sol-gel method. The morphology and structure of prepared $ErMnO_3$ were characterized by TG-DSC, XRD, FT-IR and SEM. Under ultraviolet light conditions, the photocatalytic activity of $ErMnO_3$ was evaluated using methyl orange as simulated sewage.

2. Experimental details

2.1 Preparation of $ErMnO_3$

$Er(NO_3)_3{\cdot}5H_2O$, manganese acetate, citric acid and urea were purchased from Aladdin reagent company. All reagents were analytical grade without further purification. In a typical process, at room temperature, the $Er(NO_3)_3{\cdot}5H_2O$ of 0.01 mol, citric acid of 0.01 mol and manganese acetate of 0.01 mol were successively dissolved into the 80 mL deionized water to get the solution under magnetic stirring, and then urea of 0.005 mol was added to the solution. After stirred for 1h at room temperature, the resulted solution was kept for 36 hours in 100 °C oven to get the xerogel. Then the xerogel was calcined at 800 °C for 12 hours in muffle furnace to prepare $ErMnO_3$ sample.

2.2 Characterization methods

$ErMnO_3$ sample was measured by the American TA-SDTQ600 thermal analyzer. The temperature range of the TG-DSC is from 30 °C to 1100 °C, and the heating rate is 10 °C/min. Phase composition of sample was performed with a XD-6 X-ray diffraction (Beijing Purkinje General Instrument Co. Ltd.) using Cu Kα radiation (λ=0.15406 nm) operated at 36 kV and 20 mA. The surface adsorbents of $ErMnO_3$ sample were analyzed by Nicolet 5700 Fourier transform infrared spectroscopy of the US Thermo Corporation, using the KBr pressing plate method, and the scanning wave number range was 4000-400 cm^{-1}. The morphology of $ErMnO_3$ sample was observed by Hitachi S-3400N scanning electron microscope of Hitachi Company with an accelerating voltage of 15 kV.

2.3 Photocatalytic activity testing.

The photocatalytic experiment was carried out in a self-assembled apparatus which included a light source of 150 W high-pressure mercury lamp. Typically, 20 mg $ErMnO_3$ powder was added to a 10 mL methyl orange aqueous solution (20 mg/L). Then, the solution was held for 10 min in the darkness to obtain the adsorption–desorption equilibrium, followed by the UV-light irradiation. After a set irradiation time, the suspension was centrifuged. The supernatant was subjected to the UV–visible absorption measurement. Finally, the concentration of the residual MO was evaluated by the absorbance at 464 nm, then finding its degradation rate by standard curve method.

3. Results and discussion

Figure.1 shows TG-DSC curve of the dried gel powders by sol-gel. There is a small endothermic peak appearing on the DSC curve at about 56 °C, and the TG curve has a weight loss of 1.9% from room temperature to 150 °C, which is the desorption of adsorbed water to absorb heat. A large exothermic peak appeared at about 350°C, and there is a weight loss of 5.2% from 150 °C to 500 °C, which is the result of combustion of residual organics such as citric acid adsorbed on the surface. There is a weight loss of 2.3% from 500 °C to 900 °C, the homologous DSC curve appeares a small exothermic peak at 790 °C, probably owing to the exothermic process of chemical reaction. According to the TG-DSC curve, the precursor products of $ErMnO_3$ was calcined at 800 °C to prepare pure $ErMnO_3$ samples.

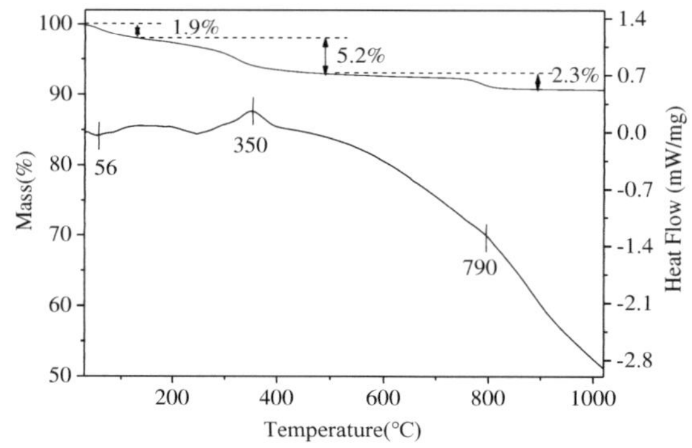

Figure.1 TG-DSC curves of the dried gel powders.

Figure.2 shows XRD pattern of the as-prepared $ErMnO_3$ sample. All peaks can be indexed to the characteristic XRD peaks of the $ErMnO_3$ according to the standard perovskite $ErMnO_3$ (PDF#14-0689) and there are no additional peaks observed, which indicates the successful synthesis of monophasic $ErMnO_3$ without external impurities.

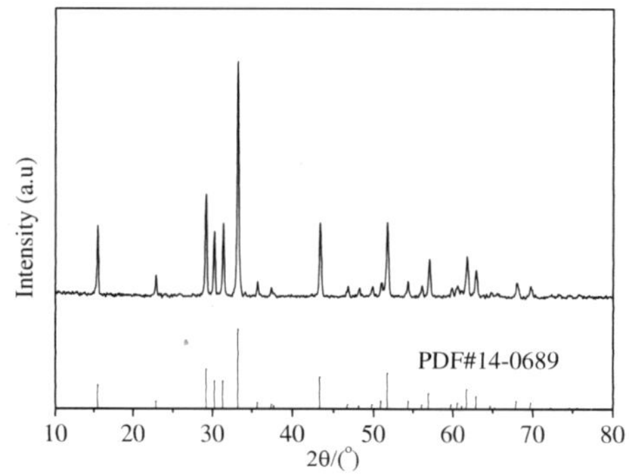

Figure.2 XRD pattern of the as-prepared $ErMnO_3$ sample.

Figure.3 is FT-IR spectrum of the as-prepared $ErMnO_3$ sample. From figure.3, there are mainly six peaks for $ErMnO_3$ sample, which are 3415 cm^{-1}, 2380 cm^{-1}, 1375 cm^{-1}, 1140 cm^{-1}, 900 cm^{-1}, 530 cm^{-1} respectively. The peaks of 1375 cm^{-1} and 3415 cm^{-1} are the -OH stretching vibration peak of sample surface absorbing water molecules. The peak around 2380 cm^{-1} is Er=O stretching vibration peak for $ErMnO_3$. The peak around 1140 cm^{-1} is Er=O torsional vibration peak for $ErMnO_3$. The peaks of 530 cm^{-1} and 900 cm^{-1} should be the vibration peak of Mn-O for $ErMnO_3$.

Figure.4 shows SEM image of the as-prepared $ErMnO_3$ sample. From figure.4, SEM image shows that the average grain size of $ErMnO_3$ sample is mainly about 100nm, the particle size distribution is relatively uniform, and there are also large particles of agglomeration. The average grain size of $ErMnO_3$ is about 100nm which has established a great foundation for the excellent photocatalytic activity of $ErMnO_3$.

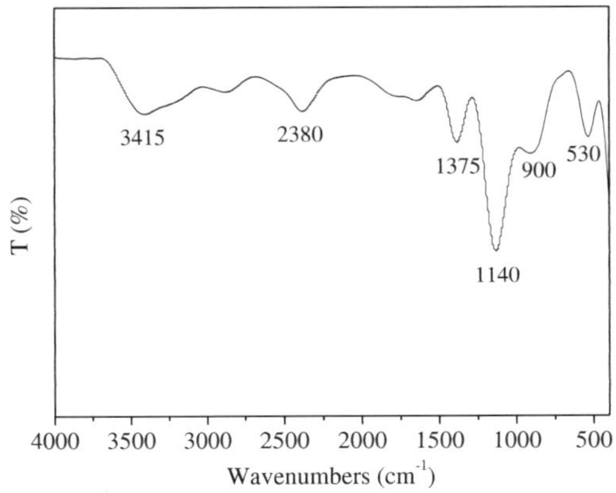

Figure.3 FT-IR spectrum of the as-prepared $ErMnO_3$ sample.

Figure.4 SEM image of the as-prepared Sample.

Figure.5 shows that photocatalytic degradation rate of methyl orange in the presence of $ErMnO_3$. From figure.5, when the illumination time was less than 60 min, the degradation rate of methyl orange increased rapidly along with the increase of illumination time. When the illumination time was more than 60 min, the degradation rate of methyl orange increased slowly along with the increase of illumination time. When the illumination time was equal to 120 min, the degradation rate of methyl orange has reached about 95%. This indicates that $ErMnO_3$ showed excellent photocatalytic activity for removal of methyl orange from simulated sewage under ultraviolet light conditions.

Figure.6 shows the kinetic curve of methyl orange degradation in presence of $ErMnO_3$. From figure.6, the degradation of methyl orange in presence $ErMnO_3$ sample was basically consistent with the first order kinetics, and its apparent rate constant is 0.022 min^{-1}.

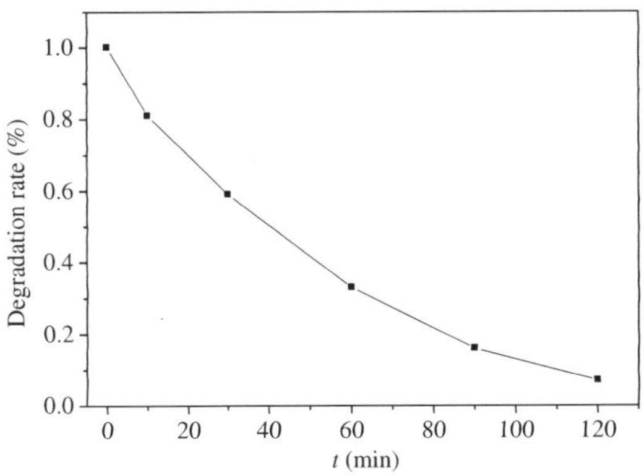

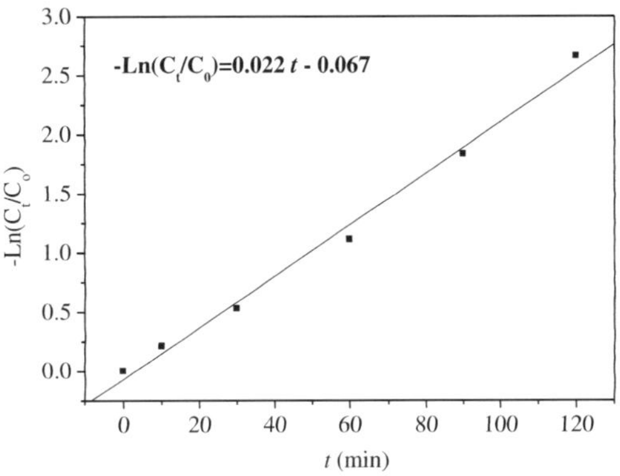

Figure.5 Photocatalytic degradation rate of methyl orange in the presence of $ErMnO_3$.

Figure.6 The kinetic curve of methyl orange degradation in presence of $ErMnO_3$.

4. Conclusions

The perovskite type $ErMnO_3$ sample was successfully prepared after calcination at 800 °C by sol-gel method using $Er(NO_3)_3$, manganese acetate, citric acid and urea as starting materials. XRD and SEM demonstrate synthesis of $ErMnO_3$ and an average grain size of 100 nm. $ErMnO_3$ shows excellent photocatalytic degradation activity for the simulated sewage methyl orange under ultraviolet light conditions. The degradation of methyl orange is basically consistent with first order kinetic model in presence $ErMnO_3$ sample, and its apparent rate constant is 0.022 min^{-1}. Therefore, $ErMnO_3$ is a kind of photocatalyst with foreground in sewage treatment.

Acknowledgements

The authors acknowledge the financial supports from National Training Programs of Innovation and Entrepreneurship for Undergraduates (No.201710347013) and General Project of Zhejiang Provincial Department of Education (Y201737320).

References

[1] Negi P, Dixit G, Agrawal H M and Srivastava R C. 2013 Structural, optical and magnetic properties of multiferroic $GdMnO_3$ nanoparticles. *Journal of Superconductivity & Novel Magnetism* **26** 1611-15.

[2] Standard E C, Stanislavchuk T, Sirenko A A, Lee N and Cheong S W. 2012 Magnons and crystal-field transitions in hexagonal $RMnO_3$ (R=Er, Tm, Yb, Lu) single crystals. *Physical Review B* **85** 2207-11.

[3] Lin-lin. 2010 Hexagonal and orthorhombic perovskite phases of $ErMnO_3$ and $TmMnO_3$ from hydrothermal systems. *Chemical Research in Chinese Universities* **26** 707-11.

[4] M Chandra Sekhar, K Padmavathi, J G Park and P Venugopal Reddy. 2003 Elastic behavior of $YMnO_3$ and $ErMnO_3$ manganites. *Modern Physics Letters B*. **17** 1119-25.

[5] Liu P, Cheng Z X, Du Y and Wang X L. 2010 Enhanced magnetic moment in $ErMnO_3$ by copper doping and negative magnetocapacitance effect. *Journal of Physics D: Applied Physics* **43** 325002-7.

[6] Romaguerabarcelay Y, Moreira J A, Almeida A, Tavares P B and Pérez D L C J. 2013 Structural, electrical and magnetic properties of magnetoelectric $GdMnO_3$ thin films prepared by sol-gel method. *Thin Solid Films* **564** 419-25.

[7] Kim Y T, Kim E S, Chae J H and Ryu J H. 2002 Ferroelectric properties of $ErMnO_3$ thin film prepared by sol-gel method. *Journal of the Korean Ceramic Society* **39** 829-34.

[8] Yan C H, Huang Y H, Wang Z M, Zhu T, Liao C S and Cheng F X. 2000 Microstructure and transport properties of sol-gel fabricated $La_{0.7}Sr_{0.3}MnO_3$ nanocrystalline films on Si(100) substrates. *Thin Solid Films* **366** 302-5.

[9] Tang P, Tong Y, Chen H, Cao F and Pan G. 2013 Microwave-assisted synthesis of nanoparticulate perovskite $LaFeO_3$ as a high active visible-light photocatalyst. *Current Applied Physics* **13** 340-43.

[10] Tang P, Hu Y, Lin T, Jiang Z and Tang C. 2014 Preparation of nanocrystalline $GdFeO_3$ by microwave method and its visible-light photocatalytic activity. *Integrated Ferroelectrics* **153** 73-8.

IOP Conf. Series: Materials Science and Engineering **362** (2018) 012022 doi:10.1088/1757-899X/362/1/012022

Cooling Effect Analysis of Suppressing Coal Spontaneous Ignition with Heat Pipe

Yaping Zhang, Shuanwei Zhang, Jianguo Wang and Gaihong Hao
College of Energy Engineering,Xi'an University of science and technology,Xi'an710054 China
E-mail:13669297513@qq.com

Key words: heat pipe;cooling effect; spontaneous ignition; thermal performance
Abstract. Suppression of spontaneous ignition of coal stockpiles was an important issue for safe utilization of coal. The large thermal energy from coal spontaneous ignition can be viewed as the latent energy source to further utilize for saving energy purpose. Heat pipe was the more promising way to diffuse effectively concentrated energy of the coal stockpile, so that retarding coal spontaneous combustion was therefore highly desirable. The cooling mechanism of the coal with heat pipe was pursued. Based on the research result, the thermal energy can be transported from the coal seam to the surface continuously with the use of heat pipe. Once installed the heat pipes will work automatically as long as the coal oxidation reaction was happened. The experiment was indicated that it can significantly spread the high temperature of the coal pile.

1. Introduction

With the acceleration of the coal exploitation, coal spontaneous combustion greatly threatens the safety of miners and properties[1]. When it was exposed to air, the coal was oxidized slowly by oxygen and generates lots of thermal energy[2]. During the process of coal extraction and transportation, the coal carbon molecule with the surrounding oxygen molecules slow oxidation exothermic reaction would also arise. This reaction can lead to spontaneous combustion with the release of toxic gases [3].

Most of studies were focused on modeling and prediction of self-heating or ignition of coal stockpiles, however, there were only a few studies devoted to prevention of spontaneous ignition of coal stockpiles[4]. Several well-known methods to suppress spontaneous ignition of coal stockpiles were periodic compaction of stockpiles[5], a low slope angle of the pile[6], pile protection by wind barriers[7] , water spray on the coal pile[8], and covering the pile with an inert layer[9].Each method had been verified to retard spontaneous ignition time in some way. However, these cooling measures had absorbed the latent heat of vaporization of water mainly in coal by heating vaporization, through hole "chimney effect" of heat to cool things down, and so this approach was affected by the influence of coal moisture[10]. In addition, coal moisture could take away vaporization latent heat in the form of water vapor. Moisture escape would increase contact area close to accelerate coal oxidation and produce large heat to strength coal fires hazard[11]. So the effect of these cooling coal fire measures was very limited.

2. Experimental Result And Discussion

2.1. The temperature distribution of coal pile with heat pipe

Fig.1 provided the comparison coal temperature distribution of the same radius range, and the same heating power with or without heat pipe,it was shown the variation curve of temperature with the heating time.It was indicated that the temperature of the coal pile was about reduced by 12°Cwith heat pipe working after 14 hours . The range of temperature drop can reach 30.1% with the effect of heat pipe, which had indicated the advantage of extract thermal energy with heat pipe.

When the radius distance from heat pipe was 50cm, as shown Fig.2,the range of temperature drop was about 10°C, and decreasing temperature was by 23.5%, and however the radius distance from heat pipe 75cm, the range of temperature drop was only around 4°C, the temperature drop was of 16.7%,as Fig.3. After 14 hours of the heat source working, it was almost the same temperature for the presence of coal pile without heat pipe.

It was indicated larger reduced temperature range was arisen in the given depth coal level for the nearer from heat pipe, and it should be noted that due to high efficiency phase change heat transfer, it can rapidly make coal pile cooling with the heat pipe operation. When heat pipe was inserted into the coal pile, the temperature rise of the coal pile was very slow. Reducing temperature was faster for heat pipe extracting thermal energy than that of the natural convection cooling.

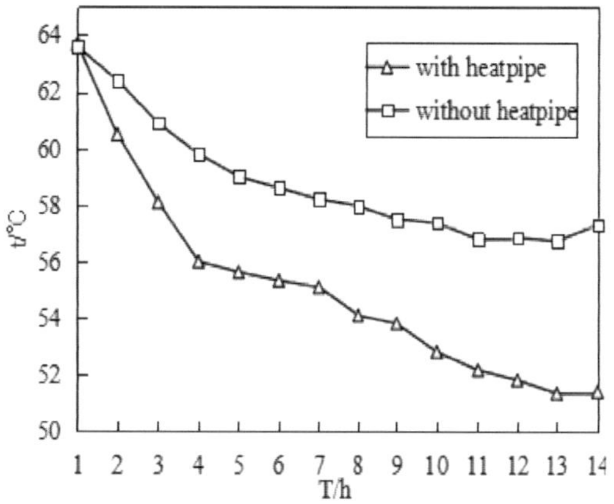

Fig 1. Temperature distribution from heat pipe radius 20cm.

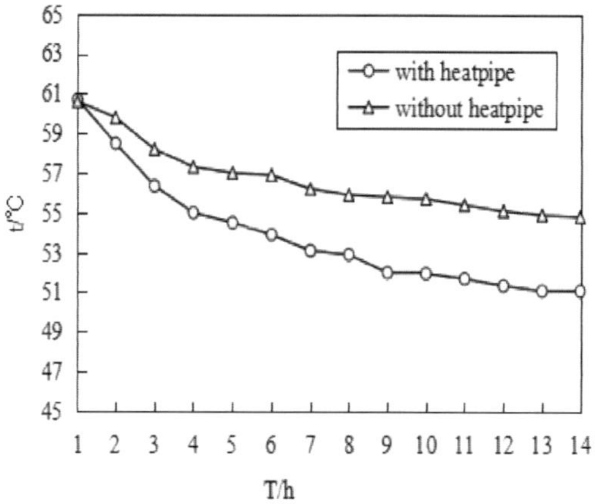

Fig 2. Temperature distribution from heat pipe radius 50cm.

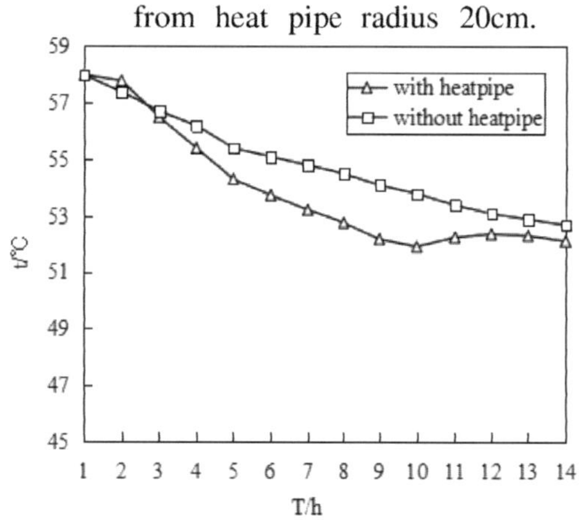

Fig 3. Temperature distribution from heat pipe radius 80cm.

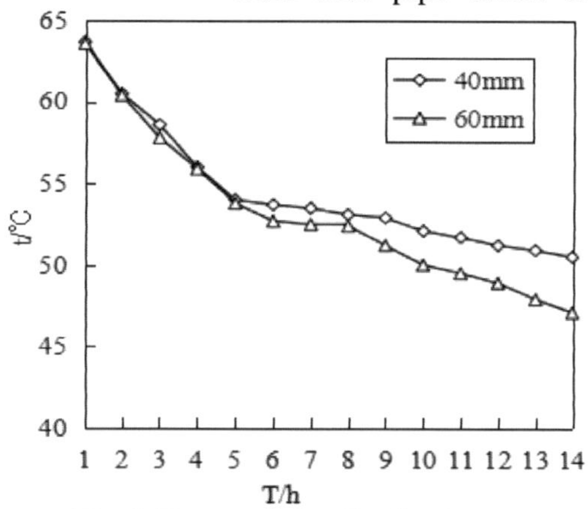

Fig 4. Temperature distribution with different depth change.

2.2. Depth change of heat pipe penetration

When the insertion depth of the heat pipe was changed from 60mm to 40mm, the temperature drop scale was reduced by 4.16% , as was shown in Fig.4.If the condensation length of the heat pipe was 30cm, the ration of condensation and evaporation section was adjusted from 1 to 2, however, the cooling effect of the heat pipe was decreased.

When it was reduced to the heat tube insertion depth of the coal pile, and thus decreasing the evaporation length, however, the length of the condensation section was increased as well. Moreover, it was decreased contact area of the thermal resource and heat pipe. Hence, the cooling effect of heat pipe was reduced on the same internal thermal source conditions.

There was a bigger cold capacity for coal depth under the 150cm, which can absorb the thermal energy of the deep coal level, and then it would make the deep level coal obviously natural cooling. However, the middle level coal was the cold source from the top and bottom, and natural temperature drop by air cooling was very slowly. After the heat pipe was inserted into the coal pile, which was served as extra central cold source. Accordingly, the range of cooling was bigger for the middle level of the natural cooling than that of the shallow coal level and deep coal level. Compared to the natural air cooling, the range of maximum temperature drop was about 13°C on the regular deep coal pile lever. When heat pipe was closer to coal level, the more obvious cooling effect for coal level with heat pipe operation.

Due to the isothermal effect of heat pipe, hence it would carry thermal energy from the cold source of the surroundings to the coal pile, the heat flow of the high temperature coal pile was circulated to the cold source of the surroundings air, and then form the temperature gradient in the fixed range of the coal pile. The research result indicated that the cooling range was more obvious in the nearer distance from the heat pipe than that of the far distance.

2.3. The influence of heat pipe inclination angle

The heat pipe layout angle was changed by original 90 ° to 60 °, the temperature of the coal pile was increased by 5.15%, Fig.5 provided temperature distribution of coal level for the depth 40cm with heat pipe, far from the heat pipe radius distance 20mm. It should be noted that due to bubble growth and movement rules as well as the common effect of flushing action for heat pipe wall, so energy transfer had some relation with the inclination angle of heat pipe.

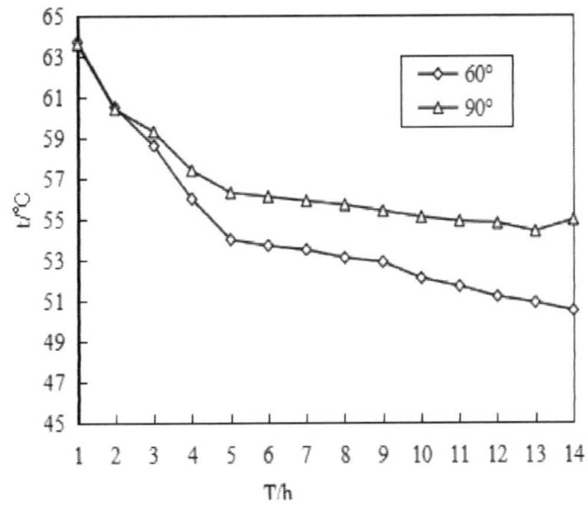

Fig 5. Temperature distribution with different angle.

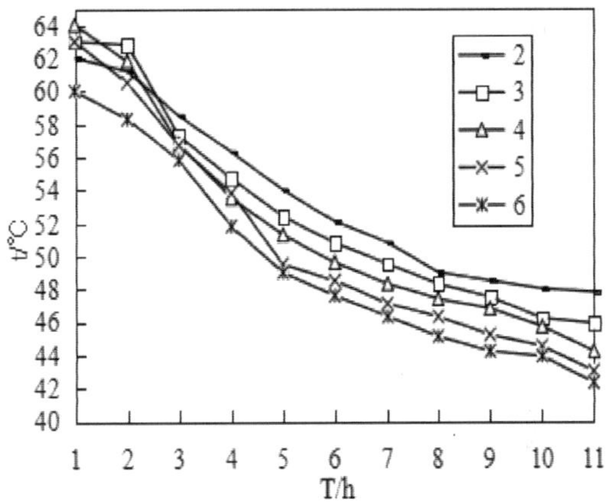

Fig 6. Temperature distribution with heat pipe number.

When the superheat of heat pipe was larger, the tube boiling was more intense for the liquid working medium. A large of the gas bubble was produced, and causing the gas disturbance in the liquid pool, some bubbles were gathered and becoming larger bubbles. Moreover, a lot of bubbles were gathered and rushed from the evaporator to condenser section by the pressure difference. When the bubbles were gathered to so big enough and reaching a certain size, which could carry some a certain amount of liquid rushing to the condensation section,which can make heat transfer power increasing under the common effect of the turbulent flow scour movement and droplets scattering, and so as to strengthening the coefficient of heat transfer.

2.4. The influence of heat pipe number

When the number of heat pipe inserted into the coal pile was different, the temperature change was shown in depth 4 cm and radius 2 cm range from heat pipe in Fig.6. The cooling effect of coal pile was more obviously with the increasing of heat pipe number. The temperature drop was more 3.5°C when the number of heat pipe was increased from 2 to 6. Moreover, the temperature of coal pile was decreased by 1.5°C when the number of heat pipe was increased from 2 to 3.Accordingly, the number of inserted heat pipe was more, the cooling effect of coal pile more better.

2.5. Heat pipe number effect of the coal lever temperature distribution

In general, the rated power was more than that of the effective heat power for the given heat pipe. So the produced cold energy was obtained by the effective power of the heat pipe.

If spontaneous combustion was occurred, the minimum temperature limitation was 60°C. In order to release this thermal energy of the coal dump, assume it was natural convection heat transfer between heat pipe and outdoor air surrounding. The accumulated thermal energy of the coal dump was calculated by the Equation(1).Only it would release these thermal energy accumulated coal dump in time, the coal temperature would be controlled less than 60°C.

$$Q=\alpha\ A(t_w-t_f) \tag{1}$$

Where, Q was thermal energy, α was natural convection heat transfer coefficient, A was heat transfer area of the heat pipe, t_W was the outside surface temperature of heat pipe, t_f was environmental temperature.The heat transfer coefficient of natural convection was derived by the basis of approximate formula.

$$\alpha = 9.77+ 0.07(t_w-t_f) \tag{2}$$

When assume $\alpha = 10.33\text{W}/(\text{m}^2\bullet\text{K})$,and then get $Q = 16.66\text{W}$. In order to make coal dump temperature control was under 60°C, about 3 roots heat pipes should be inserted in the coal dump.

Because of the definite heat pipe heat flow density was 28W/m^3, only the space of heat pipe was matched with the power of heat pipe, heat quantity of coal pile could be extracted from the underground so as to reuse secondly. If heat pipe had a plenty of power, the effect of coal cooling was more prominent for the deeper coal field and insure the safety exploitation.

3. Result

The application of heat pipe to support extinguishing coal fire, to prevent coal spontaneous as well for energy recovery purposes was thinkable.On the basis of experimental research result, it was concluded the different factors influencing the performance of coal stockpile cooling, and analyzed how these factors affect the heat transfer characteristics, some measures were proposed to strengthen thermal exchange performance of the heat pipe. With the effect of heat pipe operation, its evaporation section length, depth, and angle would also affect the temperature distribution of the coal level.The mechanism of heat pipe diffused the concentrated thermal energy for coal oxidation was pursued according to the experimental result.

The research idea for the extraction of coal fire combustion heat had provided a new way of thinking about the thermal energy reuse. Research result was shown that heat pipe could prevent and control the coal spontaneous combustion effectively, and availably diffuse heat accumulation inside the coal pile inducing temperature rise, which it had provided a new idea to suppress coal spontaneous ignition and the concentrated thermal energy extraction for coal pile.

Acknowledgment

The financial was partially supported by National Natural Science Foundation Funded Project (51504188) and It was also partially supported by Shaanxi Province Natural Science Foundation of China(2014JQ7277) ,which were greatly acknowledged.

References

[1] Chul Jin Kim,Chae Hoon Sohn, A novel method to suppress spontaneous ignition of coal stockpiles in a coal storage yard, *Fuel Processing Technology* 100 (2012) 73–83.

[2] J.H. Kim, The final report for prevention experiment of spontaneous combustion in sub-bituminous coal stockpiles, *Korea East–West Power Corporation, Dangjin Thermoelectric Power Headquarters*, 2004.

[3] V. Fierro, J.L. Miranda, C. Romero, J.M.Andres, A. Arriaga,D. Schmal, *Model predictions and experimental results on self-heating prevention of stockpiled coal, Fuel* 80 (2001),125–134.

[4] Schmidt R, Iyengar M, Steffes J, Lund V. Co-generation- Grid Independent Power and Cooling for a Data Center, *Proceedings of the ASME 2009 InterPACK Conference, IPACK*2009, July 19-23, 2009, San Francisco, USA.

[5] Zhang Qiang, Song Lianchun, Huang Ronghui, et al. Characteristics of hydrologic transfer between soil and atmosphere over near at the end of summer [J]. *Advances in Atmospheric Sciences,* 2003,20(3): 442 -452.

[6] H. Park, A.S. Rangwala, N.A. Dembsey, A means to estimate thermal and kinetic parameters of coal dust layer from hot surface ignition tests, *Journal of Hazardous Materials* 168 (2009) 145–155.

[7] S.J. Zarrouk, M.J. O'sullivan, J.D. St George, Modelling the spontaneous combustion of coal: the adiabatic testing procedure, Combustion Theory and Modelling 10(2006), 907–926.

[8] Hopkins R,Faghri. A Flat miniature heatpipeswith micro capillary grooves[J]. *ASM J. Heat Transfer*, 1999, 121: 102 -109.

[9] Salem A Said, Bilal AAkssh. Experimental performance of a heat pipe[J]. *International Journal of Heat and Mass Transfer*,1999, 26(5): 679 -684.

[10] Prabu V, Jayanti S. Integration of underground coal gasification with a solid oxide fuel cell system for clean coal utilization. *Int J Hydrogen Energy*,2012,37(2):1677–88.

[11] Chiasson,A.d., Yavuzturk,C., Walrath,D.e, EVALUTION OF ELECTRICITY GENERATION from underground coal fires and waste banks *Journal of energy resources technology*,2007,129 (2), 81-89.

[12]M.Schmid,Suhendra,H.Ruter.Heat pipes-suitable for extinguishing underground coalfires.433-437

[13] Chiasson,A.d., Yavuzturk,C. Modeling the viability of under ground coal fires as heat source for electrical power generation.*Proceeding,*Thirtieth Workshop on Geothermal Reservoir Engineering,Stanford,California.2005.

IOP Conf. Series: Materials Science and Engineering **362** (2018) 012023 doi:10.1088/1757-899X/362/1/012023

Proficiency Testing for Determination of Water Content in Toluene of Chemical Reagents by iteration robust statistic technique

Hao Wang[1], Qunwei Wang[1] and Ming He[2]

[1]Ningbo Entry-Exit Inspection & Quarantine Bureau,Zhejiang,China
[2]Zhoushan Entry-Exit Inspection & Quarantine Bureau,Zhejiang,China
wanghao@nbciq.gov.cn, wangqw@nbciq.gov.cn, hm@zs.ziq.gov.cn

Abstract. In order to investigate and improve the level of detection technology of water content in liquid chemical reagents of domestic laboratories, proficiency testing provider PT0031 (CNAS) has organized proficiency testing program of water content in toluene, 48 laboratories from 18 provinces/cities/municipals took part in the PT. This paper introduces the implementation process of proficiency testing for determination of water content in toluene, including sample preparation, homogeneity and stability test, the results of statistics of iteration robust statistic technique and analysis, summarized and analyzed those of the different test standards which are widely used in the laboratories, put forward the technological suggestions for the improvement of the test quality of water content. Satisfactory results were obtained by 43 laboratories, amounting to 89.6% of the total participating laboratories.

1. Introduction

The moisture content is common chemical reagent in the detection of the quality, is an important part of chemical reagent purity determination, influence quality, retention time and stability of chemical reagents, the significance of for the use of chemical reagents next[1]. "GB/T 690-2008", "Chemical reagent benzene", "GB/T 684-1999", "Chemical reagent toluene" and "GB/T 16494-2013" "Chemical reagent xylene" and other specifications and standards have made the requirements of moisture content. The liquid chemical reagents commonly used methods for the determination of water content by Karl Fehu titration [2], related standards are mainly GB/T 606-2003 " Chemical reagent--General method for the determination of water--Karl Fischer method " , GB/T 6283-2008"Chemical products—Determination of water Karl·fischer method(general method)". At present the verification activities of the determination of liquid chemical reagent water content is less, and the standard test results of precision requirements, the determination of toluene to verify the ability of moisture content as an example, differences in the recognition of water between the determination results of the laboratory, the inter laboratory increase the comparability of measurement results, improve the detection level laboratory [3]. Through this program, each participating laboratory can not only understand the detection ability of self, can find the gap between the laboratory and similar, but also promote the laboratory to further improve the detection ability and the level of laboratory management, at the same time for CNCA and CNAS in the field of laboratory accreditation supervision.

2. Conceptual design

A. Test Sample

IOP Conf. Series: Materials Science and Engineering **362** (2018) 012023 doi:10.1088/1757-899X/362/1/012023

(1)Sample Design: The test sample is a bottle of chemical reagent sealed in brown glass sealed bottle about 30ml of toluene. Each laboratory will receive a sample of toluene, which can be tested directly.
(2)Sample Preparation and Release: Buy 4L bottled commercial chromatography grade chemical reagent toluene from TEDIA company, shake it, and then pack it into 110 bottles of sample. Containers used are brown glass bottles of 35ml, each sample of approximately 30ml.
A brown glass bottle with a sample is wrapped in a bubble film and placed in a tin container. It is sealed and delivered to each laboratory by courier. To participate in the laboratory, at the same time receive the "work instructions", "receipt of the items to receive state confirmation form", "results report." The received samples shall be kept in sealed condition at normal temperature.
(3)Homogeneity Test: From the prepared samples, 10 bottles of samples were randomly selected for homogeneity detection. Water content was detected according to GB/T 606-2003, a general method of chemical reagent moisture determination, Carle Fischer method. The CNAS-GL03:2006" Guidance on Evaluating the Homogeneity and Stability of Samples Used for Proficiency Testing " the provisions of the single factor analysis of variance (F test) and Ss = 0.3, criterion for statistical processing of test results.
Stability Test: 6 samples were randomly selected for
(4) stability test. According to the CNAS-GL03 t test method for assessing the stability of samples, the results of each test and stability test of homogeneity data were compared, if t is less than the significance level a (usually α=0.05) degree of freedom (n1+n2-2) the critical value of Tα(n1+n2-2), the average value between the average and the stability test of homogeneity there was no significant difference, the sample is stable.
B. Test Method
This capability certification is recommended in accordance with GB/T 606-2003 or other technical equivalent or equivalent standards.
C. Statistical Evaluation Method
Check the results of this capacity verification plan laboratory and press the Z value:

$$z = (x - X)/\sigma \quad (1)$$

Type: - laboratory test results;
- specified value;
- ability to evaluate standard deviations (variability measures, target standard deviations).

The ability to verify the plan in accordance with the ISO 13528:2015 (E) " Statistical methods for use in proficiency testing by interlaboratory comparison", according to ISO 13528:2015 (E) C appendix A computing robust average algorithm to estimate the sample population mean, the average value is specified as a robust, robust standard errors as evaluation standard poor. The main process of the algorithm A is as follows. The P lab data is sorted first in ascending order:
The robust mean values and robust standard deviations for these data are calculated as. The initial value of the calculation is as follows (MED means median value):

$$x^* = \text{med } x_i \text{ (i=1,2,...,p)} \quad (2)$$

$$s^* = 1.483 \times \text{med } | x_i - x^* | \text{ (i=1,2,...,p)} \quad (3)$$

According to the following steps, update and value and calculate:

$$\delta = 1.5 s^* \quad (4)$$

For each (i=1,2),... (P), calculate

$$x_i^* = \begin{cases} x^* - \delta, \text{if } x_i < x^* - \delta \\ x^* + \delta, ifx_i < x^* - \delta \\ x_i, other \end{cases} \quad (5)$$

Next, the new sum is calculated:

$$x^* = \sum x_i^* / p \quad (6)$$

$$s^* = 1.134\sqrt{\sum\left(x_i^* - x^*\right)^2 /(p-1)} \quad (7)$$

So far, the first iteration was completed. The second iteration type (4) from beginning to end the formula (7), in turn, until the new robust average and robust convergence to the standard deviation of third decimal digits remained unchanged, and finally get the data robust average and robust standard deviation [4]. The standard uncertainty of the robust mean is calculated by the press:

$$u_x = 1.25 \times s^* / \sqrt{p} \quad (8)$$

Ability to validate the significance of each statistic and related computational methods see CNAS-GL02:2014," Guidance on Statistic Treatment of Proficiency Testing Results and Performance Evaluation ", and ISO 13528:2015(E) standards. The laboratory evaluation results in Z, the score is less than 2, the result is normal, indicating the detection ability of laboratory at the present situation is still satisfactory; 2<Z <3, the suspicious value description of laboratory current detection capability of the Z; more than 3, the outlier the current value that laboratory detection ability not satisfied with the situation.

3. Results and discussion

A. Participation in Laboratory Distribution

The ability to verify a total of 48 laboratories participated in, including production enterprises, quality supervision departments, inspection and quarantine system, social detection mechanism and a small amount of Research Institute, Petro China and other units, all submitted test results. To report the results of the laboratory are distributed in 18 provinces, autonomous regions and municipalities directly under the central government; in the laboratory the number and distribution of wide coverage, to a certain extent can reflect the ability of determination of moisture content of liquid chemical reagents in our laboratory, and provides a basis for determining and monitoring of our laboratory the detection ability.

B. Uniformity and Stability of the Sample

According to the method of uniformity in the 2.3 section in the same conditions for 2 times were calculated between mean square and mean square in the sample were 50.24, 41.27, F-measure computing statistics for 1.21, less than the critical value F0.05 (9,10) (3.02); the calculation of statistics value is 2.12, less than the critical value of 0.3 sigma (10.7). It is showed that the sample was homogeneous at 0.05 significance levels.

Stability is calculated by 2.4 in section S before =6.97, S after =8.09, and the calculated statistic t=1.64 is less than the critical value t0.05/2 (10) =2.23. It showed that the sample was stable at 0.05 significance level.

C. Test Results and Mathematical Statistics

(1)Test Result: Test results were presented in 48 laboratories, and the results were shown in table 1.

Table 1. Test result of water content

No.	Result	No.	Result	No.	Result	No.	Result	No.	Result
1	364.0	11	455.6	21	357.0	31	409.0	41	408.7
2	365.0	12	394.6	22	353.4	32	310.6	42	429.5
3	390.8	13	337.4	23	391.7	33	395.0	43	449.5
4	366.2	14	398.0	24	370.1	34	432.1	44	454.4
5	374.2	15	428.4	25	533.4	35	400.0	45	717.0
6	409.6	16	400.0	26	412.2	36	368.0	46	402.2
7	391.1	17	405	27	371.2	37	350.0	47	411.0
8	404.8	18	469.5	28	344.5	38	401.3	48	357.0
9	404.3	19	402.4	29	375.6	39	300.5		
10	397.0	20	399.6	30	415.1	40	407.3		

Statistics the frequency of all laboratory results, grouped according to the content, and frequency histograms, as shown in figure 1.

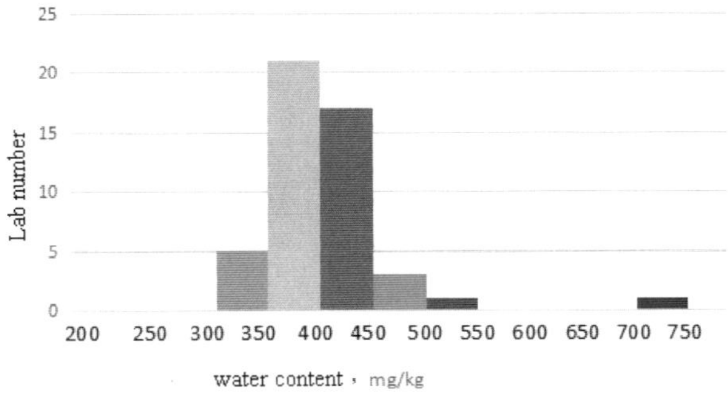

Figure 1. Histogram showing of frequency distribution of testing results.

Frequency distribution diagram shows that although there are individual extreme values, the overall data of this capability verification result appear to be in the middle and less at both ends.

Mathematical Statistics of Test Results: The results show that the ability of statistical validation plan, there are 43 laboratories for the detection results with satisfactory results, accounting for 89.6% of the total number of detection results; 3 laboratories results for the problem, accounting for 6.2% of the total number of detection results; 2 laboratories were not satisfied with the results, accounting for 4.2% of the total number of feedback results. For statistical parameters, see table 2.

Table 2. Test result of water content

Test item	Water ontent
Result number	48
Robust mean（mg/kg）	395.3
Robust standard deviation（mg/kg）	35.7
Standard uncertainty（mg/kg）	6.4
Robust coefficient of variation（%）	9.0
Minimum（mg/kg）	300.5
Maximum（mg/kg）	717.0
Range（mg/kg）	416.5

In order to clearly indicate the results of each laboratory's ability validation program, the Z radio are arranged in order of size (Fig. 2). Each column is marked with the code of the laboratory. From the bar chart, it is easy for each laboratory to compare its results with the results of other participating laboratories to find out where the results are in the plan.

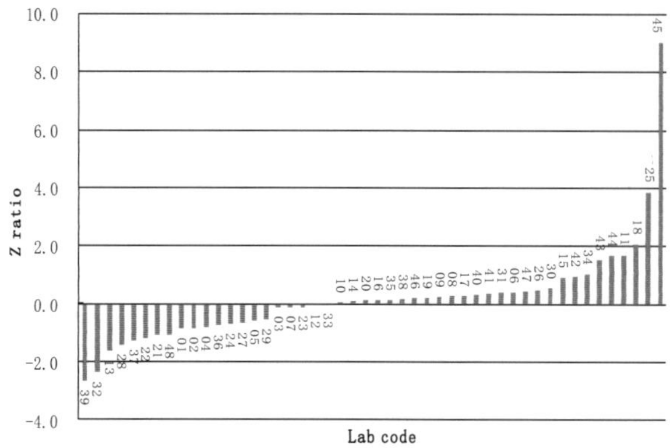

Figure 2. The Histograms of Z ratio of water content Technical Analysis

In the laboratory the ability to verify the determination of water content standards, including GB/T 606-2003 "Chemical reagent--General method for the determination of water--Karl Fischer method", GB/T 1600-2001"Testing method of water in pesticides"," GB/T 2288-2008 "Coking products - Determination of moisture content", GB/T 6283-2008"Chemical products—Determination of water Karl·fischer method(general method) ", SH/T 0246-1992 "Light petroleum products - Determination of moisture content (power law) "and ASTM D1364-02 (2012) " Standard Test Method for Water in Volatile Solvents (Karl Fischer Reagent Titration Method) ", ASTM D6304-16e1 " Standard Test Method for Determination of Water in Petroleum Products, Lubricating Oils, and Additives by Coulometric Karl Fischer Titration" , ASTM E203-16 " Standard Test Method for Water Using Volumetric Karl Fischer Titration", and ASTM E1064-16 "Standard Test Method for Water in Organic Liquids by Coulometric Karl Fischer Titration", determination of the above criteria are based on the principle of Carle Fischer method, there are 70.8% laboratories using GB/T 606-2003 and GB/T 6283-2008 two test standard; 71.4% laboratories using the moisture meter of Metrohm and METTLER TOLEDO companies.

For the determination of water content of chemical reagent, there are many factors that influence the result of determination. The following points should be paid attention to in every laboratory:

1) the state of the instrument is the key factor for the result to be accurate, to ensure the tightness of the instrument line, to isolate the reaction system from the outside, and to avoid the pollution of the platinum electrode.

2) it should be minimized the water content of the Carle Fischer Reagent itself and the instrument system[5].

3) the sampling volume is correct or not is the key, the minimum amount of sampling to ensure its moisture content should be greater than the resolution of the instrument, sampling the maximum amount guarantee water capacity of the water does not exceed the system, such as the Kulun general Carle Fischer method, sample volume control in the absolute amount of water can get accurate results in 100μg -1000μg, and Carle Fischer method considering Carle Fischer Reagent Titration, besides, used to determine whether the balance after verification[6].

4) the automatic instrument parameters should be adjusted to a more suitable state, such as drift should be less than a certain value, otherwise it will bring greater uncertainty.

5) recommend using Merck, kGaA, apura, water-standard, Hydranal[7].

4. Conclusions

This work explores the use of toluene samples to carry out the ability to detect chemical reagent liquid water content verification, 48 laboratories in 18 provinces, autonomous regions and municipalities directly under the central government to participate in the verification activities this ability, through the statistical A algorithm based on robust technology, the test results of the satisfaction rate is 89.6%, that robust coefficient of variation was 9%. Most of the domestic testing laboratory has ability to detect liquid chemical reagent water, and objectively reflect the level of detection in the laboratory,

laboratory technology and promote the level of laboratory management and other aspects of the improvement, in particular are not satisfied with the results of detection in the validation plan ability of the laboratory, should actively look for reasons daily, to take steps to ensure the accuracy of test results.

5. Acknowledgements

General Administration of Quality
Supervision, Inspection and Quarantine（2016IK175）.

6. References

[1] Rong Meng and Ru-Tian Shang 2001 Chemical Reagents. **1** 39-41
[2] Claudia C and Dongsheng Bu 2011 Anal. Chim. *J. Acta* **1-2** 84-93
[3] Qi-Bing YU and Di SU 2014 Chinese J.Anal.Chem **9** 1362-1368
[4] Xiao-Lu XIN and Tao LIU 2016 Physical Testing and Chemical Analysis(Part B: Chemical Analysis) **52** 819-824.
[5] Yiu-chung Yip and Siu-kay Wong 2011 Trends Anal.Chem **4** 628-640.
[6] YAZGAN S and BERNREUTHER A. D 2006 Food Cshem **3** 411-417.
[7] Hai-Feng WANG and Kang MA 2011 Chemical Reagents **8** 711-714.

IOP Conf. Series: Materials Science and Engineering **362** (2018) 012024 doi:10.1088/1757-899X/362/1/012024

Prescription Proportion of Pomegranate Extract Gallic Acid Gel by Orthogonal Design

Gaofu Fan[1], Xiushu Liu[1], Jie Tang[1], Jumei Gong[1], Entao Fu[1],Yuhua Cai[1] and Zhenguo Xu[1,*]

[1]Faculty of Bioscience Engineering, Hefei technology college, Hefei, China
chzyjsxy2006@126.com

Abstract. The aim of the present work was to optimize the formulation of pomegranate extract gallic acid gel by orthogonal design. Using orthogonal design, propylene glycol, carbomer-940 and gel pH level as influencing factors, the evaluation key index was external apearance malleability, uniformity, and eccentric for gel, and the optimum formula was selected. The present findings suggest that 10% propylene glycol, 1.5% Carbopol-940, and gel pH in the range of 4.5~5.5, and the indexes of the optimal. The inclusion complexes showed that after the orthogonal design, the preparation process was simple, stable and controllable quality, with production feasibility.

1. Introduction

Pomegranate was one of the special fruits of Huaiyuan area in Bengbu, Anhui Province. It had the functions of food and medicine and health care. It could prevent and treat malignant tumors, infectious diseases and cardiovascular diseases. These effects were derived from polyphenols, riched in pomegranate fruit, and gallic acid was one of the most simple components of polyphenols, and had anti-inflammatory, anti-oxidation, anti-bacterial, anti-virus, anti-allergic and anti-effect and other effects [1-5], especially with the skin Related anti-inflammatory anti-bacterial itching, anti-allergy, wound repair, whitening and other functions [6-8], to provide a theoretical basis for the clinical development of external preparations.

Gels was a more commonly used topical preparations, in the preparation process, need to consider prescription technology and raw materials base, and the preparation of bioavailability is not high. According to the clinical application characteristics, need to consider the physical and chemical properties of pomegranate extract gallic acid, matrix components and skin physiological characteristics, using orthogonal design, propylene glycol, carbomer-940 and gel pH level as influencing factors, the evaluation key index was external apearance malleability, uniformity, and eccentric for gel, and the optimum formula was selected to research and develop external use of new formulations of gels.

2. Material

2.1. Materials and Reagents

Pomegranate peel (Anhui Bengbu Huaiyuan area); gallic acid extract (self-prepared); medicinal grade accessories Kaibuomu -940 (Beijing people Yikang Technology Co., Ltd.); use test analysis level triethanolamine, propylene glycol.

2.2. Instruments and Equipment

RE-2000A rotary evaporator (Shanghai Yarong biochemical instrument factory), SCIENTZ-10ND multi-manifold gland type freeze dryer (Ningbo Xinzhi Biotechnology Co., Ltd.), HH-W0-20 intelligent constant temperature oil bath pot (Shenzhen Sanli Chemicals Co., Ltd.), PHG-3C desktop acidity meter (Shanghai instrument and electrical scientific instrument Limited by Share Ltd.), DHG-9075A desktop electric heating oven (Shanghai Qi Xin Scientific Instrument Co., Ltd.), FA1004 type of electronic Analytical balance (Tianjin Tianma Hengji Instrument Co., Ltd.), TGL-20B high-speed desktop centrifuge (Beijing Jiayuan Industrial Technology Co. Ltd.).

3. Methods

3.1. Preparation of Ellagic Acid Extract

The pomegranate peels in Huaiyuan area of Bengbu, Northern Anhui Province were dried and crushed , and the alcohol was extracted with different concentrations of ethanol (concentration: 90%, 70 % and 40%) for 1.5h. The extracts were filtered and concentrated under reduced pressure. Ethanol was recovered and the extract was concentrated and concentrated to a solution of 0.5 g of pomegranate per ml of liquid. Ultrasonic combination of acid hydrolysis method to optimize the extraction, D201 macroporous resin adsorption technology separation and purification of pomegranate extract gallic acid, the separation and purification of gallic acid by freeze-drying was placed in reserve.

3.2. Orthogonal Design

In this study, the propylene glycol, carbomer concentration and pH value were set as the influencing factors of the gel preparation. The experimental design was carried out using L9 (34) orthogonal test table, and 10%, 20% and 30% propylene glycol; 1.0%, 1.5% and 2.0% Carbopol-940; pH is set at 4.5 - 5.5, 5.6 - 6.5 and 6.6 - 7.5 as the table of nine factors were prepared corresponding gel preparation, see Table 1.

Table1. Factors table

level	factors		
	propylene glycol(A)%	carbomer -940(B)%	pH value(C)
1	10	1	4.5-5.5
2	20	1.5	5.6-6.5
3	30	2	6.6-7.5

3.3. Gel inspection Standards

The appearance, coating property, uniformity and centrifugation of the gel were examined, and the appearance of the gel was measured by the gelatin surface, and the appearance shape was evaluated by smoothness. The gel was coated with a glass rod, If the coating on the coating evenly distributed, and no adhesion on the glass rod was good for the coating; whether the gel was fine and even, and had no granular micelle, caking exists to evaluate its uniformity; respectively, the group of gel to 3000 r/Min speed for 30 min centrifugal, to observe whether there was stratification to determine its centrifugal traits. Each indicator 0 ~ 2.5 points, 10 was divided into the highest value, see Table 2 [9-11].

Table2. Scoring criteria list

index	scores			
	Excellent (2.0-2.5)	Good(1.0-1.5)	Medium(0.8-0.9)	Poor(0-0.2)
external appearance	smooth surface	smooth surface finish	dull surface	Having a surface with no gloss
malleability	easy	can, but less delicate	can, but the fineness is poor	The outflow of liquid or viscous difference
uniformity	fine and smooth	lower delicate	low ammonia	coarseness
eccentric	the appearance of the same	no layered	having particles or agglomerates on the surface	layered

3.4. Statistical analysis Method
This study was analyzed by SPSS19.0 statistical software package. The data were analyzed by single factor analysis. The difference was statistically significant by P <0.05.

4. Results

4.1. The results of Orthogonal Test are Shown in Table 3

Table 3. The results of orthogonal experiment (n=3)

factors	A	B	C	D	results
1	1	1	1	1	9.2
2	1	2	2	2	9.5
3	1	3	3	3	6.5
4	2	1	2	3	8
5	2	2	3	1	7.5
6	2	3	1	2	8.3
7	3	1	3	2	6.3
8	3	2	1	3	9
9	3	3	2	1	6.3
K_1	8.4	7.833	8.833	7.667	
K_2	7.933	8.667	7.933	8.033	
K_3	7.2	7.033	6.767	7.833	
R	1.2	1.634	2.066	0.366	

4.2. Analysis of Variance Results
Through the variance analysis, the three main factors of the study, there was significant B and C factors (P < 0.05), but there is no obvious correlation between the factors and the effect of A (P > 0.05), See Table 4 for details.

4.3. Verification Experiment
Combination with the results obtained in Table 3 and Table 4, it was concluded that the effect of pH value (C) was the highest in the three influencing factors of this study, and that of the adjuvant material carbomer-940 was the second, and propylene glycol (B) did not show correlation, and with the increase in the number of its effect weakened. From the industrial production to reduce energy consumption and control costs and many other considerations, $A_1B_2C_1$ combination of the best, the specific process to take 1.5 g carbomer-940, add appropriate amount of distilled water, put it overnight, until it was fully expanded, weighed 6g pomegranate extract gallic acid and added a certain amount of propylene glycol to dissolve it until it was dissolved and then added to the gel matrix, and to adjust its pH by triethanolamine, adjusted to 4.5 ~ 5.5 stop, then add distilled water to 100 g ,well mixed. According to this method, three parallel preparations, showed that the preparation process was highly stable and the operation was simple. The detailed results were shown in table 5.

Table 4. Analysis of variance (n=3)

soruces of variation	sum of squars	n	MS	F	Significance difference
A	2.2	2	1.1	10.86	>0.05
B	4	2	2	19.79	<0.05
C	6.44	2	3.22	31.86	<0.05
D	0.2	2	0.1		
error	0.2				
total variation	13.91	8			

$F_{0.05}(2, 2)=19.00$

Table5. Parallel research results

No	overall score	χ2	P value
1	9.8	15.779	<0.05
2	10		
3	9.9		
average	9.9		

5. Discussions

In the 2015 edition of the "Chinese Pharmacopoeia" four appendixes has been defined in the gel, refers to the amount of matrix and extract prepared by the gel characteristics of the thick or semi-solid preparation, the quality requirement of gel should be uniform and delicate, and keep the colloid at room temperature, not dry or liquefied. As the preparation of this study was mainly used for skin smear, taking into account its absorption characteristics, so in the preparation of gels in the process, the main inspection for the preparation were the malleability, stability, appearance, uniformity and other items.

Carbomer 900 series was a high polymer acrylic monomer and allyl pentaerythritol allyl sucrose or crosslinked, acidic groups in molecular structure containing 52% to 68%, its 1% aqueous solution of pH is 2.5 ~ 3, adding alkali (sodium bicarbonate) or organic alkali (such as triethanolamine) neutralization. The hydrogels formed after neutralization was stable in the range of pH4.0 to 11, and the viscosity was the most viscous in the range of pH 6.5 to 7.5. When the pH is less than 3 or greater than 12, the viscosity decreased [12]. Carbomer after neutralization had good adhesion, gelling property, but also had good emulsifying properties, thickening, suspending and film-forming properties, widely used in emulsion, cream, and gel preparation. At present, the major domestic carbomer models have 934, 940, and 941. The carbomer 940 was easy to spread, good adhesion, easily cleaning, no greasy feeling, could absorb the tissue exudates, no stimulation of the skin, often used as a gel matrix, was widely used in the mucosa, transdermal drug delivery preparation.

In this study, we found that the pH of the gel and the gel matrix Carbomer-940 were directly related to the effect of the gel formulation during the preparation of the pomegranate extract gallic acid gel preparation ($P < 0.05$) The reason for this phenomenon may be that the pH value directly affects the solubility of pomegranate extract gallic acid in propylene glycol, Its effect was related to the formation of hydrogen bonds between hydroxyl groups and hydroxyl groups of hydroxyl radical in pomegranate extract gallic acid, and the concentration of carbomer-940 was related to the molding of the gel.

6. Conclusions

The product quality inspection found that, at a rate of 3000r/min centrifugal 30min, and then put it into the oven in 24h, no stratification occurred. The results showed that the pomegranate extract gallic acid gel prepared by using the concentration of 10% propylene glycol as the solubilizer, 1.5% carbomer-940 as the gel matrix and controlling the pH of the gel between 4.5 and 5.5, had a very good stability. It was worthy of wide industrial production test further.

7. Acknowledgements

This study was supported by the Key project of Anhui provincial excellent young talent support program (gxydZD2017128); Natural Science Foundation of Anhui Province Key Project (KJ2015A440, KJ2016A616); Key quality engineering project of Anhui Provincial education Department(2014sxzx047,2016msgzs062,2015ckjh157). Gaofu Fan was responsible for the design and writing of the thesis; Zhenguo XU and Xiushu LIU put forward constructive suggestions on this paper and made some modifications; Jie Tang work for part of the experimental; Jumei Gong , Entao Fu and Yuhua Cai are responsible for the literature review of the article. And the authors declare that they have no conflicts of interest.

8. References

[1] Zheng X.H, Yang J and Yang Y 2017 Research progress on pharmacological effects of gallic acid. *J. Chin Hosp Pharm* **1** 94-99

[2] BenSaad L.A, Kim K.H, Quah C.C, Kim W.R and Shahimi M 2017 Anti-inflammatory potential of ellagic acid, gallic acid and punicalagin A&B isolated from Punica granatum. *J. BMC Complementary and Alternative Medicine* **2** 15-18

[3] Bhattacharyya S, Ahammed S.M, Saha B.P and Mukherjee P.K 2013 The gallic acid–phospholipid complex improved the antioxidant potential of gallic acid by enhancing its bioavailability. *J. AAPS Pharm Sci Tech* **3** 1025-1033

[4] Subramanian A.P, Jaganathan S.K, Mandal M, Supriyanto E and Muhamad I.I 2016 Gallic acid induced apoptotic events in HCT-15 colon cancer. *J. World J Gastroenterol* **15** 3952-3961

[5] Russell Jr L.H, Mazzio E, Badisa R.B, ZhiPing Zhu, Agharahimi M, Oriaku E.T and Carl B.G 2012 Autoxidation of gallic acid induces ROS-dependant death in human prostate cancer LNCaP Cells. *J. Anticancer Res* **5** 1595–1602

[6] Fawole O.A, Makunga N.P and Opara U.L 2012 Antibacterial, antioxidant and tyrosinase-inhibition activities of pomegranate fruit peel methanolic extract. *J. BMC Complementary and Alternative Medicine* **2** 31-35

[7] Su T.R, Lin J.J, Tsai C.C, Huang T.K, Yang Z.Y, Wu M.O, Zheng Y.Q, Su C.C and Wu Y.J 2013 Inhibition of elanogenesis by Gallic Acid. *J. Possible Involvement* **4** 43-45

[8] Meritxell M, Cristina A, Vanessa M, Manel L, Alfons D, Parra J.L and Luisa C 2013 Cosmetotextiles with gallic acid: skin reservoir effect. *J. Journal of Drug Delivery* **1** 7

[9] Luo Y.Q, Ablizl A, Zhong L.H and Aisa H 2013 Formula of Punica granatum peel phenols gel by orthogonal design. *J. Xinjiang Med Univ* **6** 734-736

[10] Fan G.F, Liu X.S, Tang J, Gong J.M and Cai Y.H 2016 Prescription ratio of pomegranate ellagic acid gel by orthogonal design. *J. Chin Mod Appl Pharm* **12** 1523-1525

[11] Yang Y.Y, Jiang X and Wang Y.Q 2017 Formulation optimization of compound tripterygium hypoglaucum gel by orthogonal test. *J. Chi Phar* **7** 961-963

[12] Guo H.Y, Yi B.W, Yan X.P, Wang C.H and Yu L 2013 Application status of carbomer in gel preparations. *J. Chinese Journal of Experimental Traditional Medical Formulae* **17** 371-374

IOP Conf. Series: Materials Science and Engineering **362** (2018) 012025 doi:10.1088/1757-899X/362/1/012025

System Analysis by Mapping a Fault-tree into a Bayesian-network

B Sheng[1], C Deng[2], Y H Wang[3] and L H Tang[1*]

1 Department of Mechanical Engineering, The University of Auckland, Auckland, New Zealand
2 School of Mechanical Science & Engineering, Huazhong University of Science & Technology, Wuhan, China
3 China Electronic Product Reliability and Environmental Testing Research Institute, Guangzhou, China

*E-mail: l.tang@auckland.ac.nz

Abstract. In view of the limitations of fault tree analysis in reliability assessment, Bayesian Network (BN) has been studied as an alternative technology. After a brief introduction to the method for mapping a Fault Tree (FT) into an equivalent BN, equations used to calculate the structure importance degree, the probability importance degree and the critical importance degree are presented. Furthermore, the correctness of these equations is proved mathematically. Combining with an aircraft landing gear's FT, an equivalent BN is developed and analysed. The results show that richer and more accurate information have been achieved through the BN method than the FT, which demonstrates that the BN is a superior technique in both reliability assessment and fault diagnosis.

1. Introduction

From the view of reasoning and fault status description, the BN method developed in recent years is better than the FT. Not only the BN method has an ability to describe the polymorphism and the non-determinism of events, but also the bottom-event is not limit to the independent faults, which is more appropriate to analyse the security and reliability of complex systems [1-3].
A large number of researches have been carried out on the transformation from the FT into the BN [4-6]. Zhou [7] proposed a BN method for calculating the minimal cut/path sets, and the importance degrees for structure, probability and critical, while the correctness of this method was not proved. Due to this shortcoming, this paper aims to prove the correctness of the BN method by verifying equations of three importance degrees described above. Then a fault case of an aircraft landing gear was processed by the BN method to study the application in the reliability assessment and fault diagnosis.

2. The Bayesian network

Definition 1: a BN is a directed acyclic graph, which is constituted by points (represent variables) and directed edges (connect these points). The directed edge is represented as a single arrow-line in the BN, and the direction of this arrow-line is from a parent point to a child point [8]. A simple BN is shown in Fig.1.
Moreover, the BN can be explained by qualitation and quantitation:

- Qualitation: the BN takes a directed acyclic graph to describe the dependence/independence relationships between the variables.
- Quantitation: the BN uses the conditional probability distribution to present the dependence relationships between the variables and their parent points. That is, each point in the network corresponds to a conditional probability table (CPT). For example, the CPT of a point X_i can be presented as $P(X_i \mid \pi(X_i))$, here $\pi(X_i)$ is the parent point set of the point X_i.

If there is a BN¥, its arbitrary variable set is $N = \{X_1 L\ X_n\}$, and the corresponding joint probability table is $\Pr(X_1 K\ X_n)$, so its joint probability distribution can be obtained by multiplying each variable with its corresponding joint probability table [9,10], that is $\Pr(X_1 X_2 L\ X_n)=\prod_{i=1}^{n} P(X_i \mid \pi(X_i))$.

A	θ_A
0	.3
1	.7

A	B	$\theta_{B\mid A}$
0	0	.1
0	1	.9
1	0	.8
1	1	.2

Figure 1. A simple BN.

2.1. Mapping method

The FT and the BN are all the expressions for systems' causal logic. The logic gate in the FT represents the determined relationship between two events, while the CPT in the BN expresses the undetermined relationship between points. Obviously, setting the value of the CPT is the key for mapping a FT into an equivalent BN [4]. A brief description of the basic mapping steps is as follows.

- Each event (bottom-event, middle-event and top-event) in a FT will be presented as a parent point or a child point in an equivalent BN, and named as the event name. Moreover, for the repeated events, there will be a same point for them in the equivalent BN [5]. Furthermore, in the equivalent BN, the top-event in the FT will be treated as the point T, and the bottom-event x_i in the FT will be treated as the point X_i which should be 0 or 1(0 means normal, 1 means fault).
- Establish connections for points in the equivalent BN according to relationships of events in the FT.
- Determine the priori probability of the bottom-point in the equivalent BN according to the failure probability of the corresponding bottom-event in the FT.
- Set the CPT of points in the equivalent BN according to the corresponding middle events and top events in the FT [7, 11].

2.2. Bi-directional reasoning

The reasoning for a BN can be summarized as: solving issues of the posterior probability of interested variables based on evidence variables and network structures. The mathematical description of a BN is as follows.

For any BN¥, the variable set N can be divided into three categories: the evidence variable $E=e$, the query variable Q (the interested variable) and the other variables $S = N \setminus (Q \cup M)$. That is, calculating the probability distribution $\Pr(Q \mid E=e)$ can solve posterior probability issues. Moreover, solutions of the probability distribution can be the same solutions of the Eq. (1) and the Eq. (2) based on the conditional independence, the chain rule and the Bayes theorem.

$$\Pr(Q, E = e) = \sum_{S} \prod_{i=1}^{n} P(X_i \mid pai(X_i)) \tag{1}$$

$$\Pr(\boldsymbol{Q} \mid \boldsymbol{E}=\boldsymbol{e})=\frac{\Pr(\boldsymbol{Q}, \boldsymbol{E}=\boldsymbol{e})}{\sum_{\boldsymbol{Q}} \Pr(\boldsymbol{Q}, \boldsymbol{E}=\boldsymbol{e})} \tag{2}$$

A BN has an ability of bi-directional reasoning, which contains a bottom-up causal reasoning and a top-down diagnostic reasoning [2, 5]. As described above, the points of the bottom-event and the top-event in the equivalent BN are X_i and T, respectively. So the bottom-up causal reasoning can be expressed as calculating the probability of the top-event when the bottom-event is fault (Eq. (3)) and the top-down diagnostic reasoning is just the reverse (Eq. (4)).

$$\Pr(T=1 \mid X_i=1)=\frac{\Pr(T=1, X_i=1)}{\Pr(X_i=1)} \tag{3}$$

$$\Pr(X_i=1 \mid T=1)=\frac{\Pr(T=1, X_i=1)}{\Pr(T=1)} \tag{4}$$

It can be seen from comparisons of the Eq. (3) and the Eq. (2), and of the Eq. (4) and the Eq. (2), the bottom-up causal reasoning and the top-down diagnostic reasoning are both belong to the posterior probability issue. This issue can be solved by reasoning algorithms such as the tree propagation algorithm, the variable elimination algorithm and the Aho–Corasick algorithm [8, 12, 13].

3. Three importance degrees

For a FT, the first step of a qualitative analysis is to get an expression of the probability function $g(q)$ of an accident tree [14, 15]. Furthermore, the occurrence probability of a top-event in a FT can be obtained through calculating the marginal probability $\Pr(T=1)$ of an equivalent BN (Eq. (5)) [5].

$$\Pr(T=1)=g(q) \tag{5}$$

Setting a FT as a $\overline{FT}$, an equivalent BN as a $\overline{BN}$ and the number of bottom-events as n. Zhou [7] proposed a $\overline{BN}$ method to compute three importance degrees (structure, probability and critical). However, he did not provide any evidence to support his method. So the following content will prove the correctness of his method based on the Eq. (5).

3.1. The Structure Importance Degree

Top-event is the structure function $\phi(x)=\phi(x_1, x_2, \ldots x_n)$ of the bottom-event. The value of $\phi(x)$ represents the value of the bottom-event x_i, which could be 0 or 1. Furthermore, the structure importance degree of the bottom-event x_i in $\overline{FT}$ and the corresponding point X_i in $\overline{BN}$ could be calculated by the Eq. (6) [14] and the Eq. (7) [7], respectively.

$$\begin{aligned} I_{\phi(i)}=\frac{1}{2^{n-1}} \sum_{k_1, k_2, \ldots, k_{i-1}, k_{i+1}, \ldots, k_n=0}^{1} & [\phi(x_i=1, x_1=k_1, \ldots, x_n=k_n) \\ & -\phi(x_i=0, x_1=k_1, \ldots, x_n=k_n)] \end{aligned} \tag{6}$$

Note in the Eq. (6), the $\sum$ means traverse the values of all bottom-events as 0 or 1(expect the bottom-event x_i).

$$I_{\phi(i)}=\Pr(T=1 \mid X_i=1)-\Pr(T=1 \mid X_i=0) \tag{7}$$

Moreover, reset the conditional probability distribution of each bottom-event as $\Pr(X_j=0)=1/2, \Pr(X_j=1)=1/2$, $j \neq i$. The proof is as follows.

The structure importance degree is not related to the failure probability of a certain bottom-event. In this work, the failure probability of the bottom-event x_j's was set as $1/2$ and $j \neq i$. So, the conditional probability distribution of the point X_j (equal to the bottom-event x_j in the FT) in the equivalent BN is $\Pr(X_j=0)=1/2$ and $\Pr(X_j=1)=1/2$.

Then the Eq. (8) and the Eq. (9) can be obtained due to the independence of $X_1, X_2, \ldots, X_n$.

$$\begin{aligned}&\Pr(X_1, X_2, \cdots, X_{i-1}, X_{i+1}, \cdots, X_n) = \\ &\Pr(X_1)\cdot\Pr(X_2)\cdot\cdots\cdot\Pr(X_{i-1})\cdot\Pr(X_i)\cdot\cdots\cdot\Pr(X_n) = \frac{1}{2^{n-1}}\end{aligned} \tag{8}$$

$$\begin{aligned}&\Pr(X_1, X_2, \cdots, X_n) = \Pr(X_1)\cdot\Pr(X_2)\cdot\cdots\cdot\Pr(X_n) \\ &= \frac{1}{2^{n-1}}\cdot\Pr(X_i)\end{aligned} \tag{9}$$

In a FT, the meaning of $\phi(x_1 = k_1, \cdots, x_n = k_n) = 0$ is that all values of bottom-events x are 0, the top-event would not occur definitely. In an equivalent BN, it means that when the status (values) of all points X are given (as 0), the probability of point T gets the status of 1 is zero. Similarly, the meaning of $\phi(x_1 = k_1, \cdots, x_n = k_n) = 1$ is that all values of bottom-events x are 1, the top-event would occur definitely. In an equivalent BN, it means that when the states of all points X are given (as 1), the probability of point T gets the status of 1 is one. So the Eq. (10) can be obtained.

$$\begin{aligned}&\phi(x_1 = k_1, \cdots, x_n = k_n) \\ &= \Pr(T = 1 \mid X_1 = k_1, \cdots, X_n = k_n)\end{aligned} \tag{10}$$

The Eq. (11) can be got after putting the Eq. (10) into the Eq. (9):

$$\begin{aligned}I_{\phi(i)} = \frac{1}{2^{n-1}} \sum_{k_1,k_2,\cdots,k_{i-1},k_{i+1},\cdots,k_n=0}^{1} &[\Pr(T = 1 \mid X_i = 1, X_1 = k_1, \cdots, X_n = k_n) \\ &- \Pr(T = 1 \mid X_i = 0, X_1 = k_1, \cdots, X_n = k_n)]\end{aligned} \tag{11}$$

Moreover, the first half of the Eq. (11) can be expressed as:

$$\begin{aligned}&\Pr(T = 1 \mid X_i = 1, X_1 = k_1, \cdots, X_n = k_n) = \\ &\frac{\Pr(T = 1, X_i = 1, X_1 = k_1, \cdots, X_n = k_n)}{\Pr(X_i = 1, X_1 = k_1, \cdots, X_n = k_n)} = \\ &\frac{\Pr(T = 1, X_i = 1, X_1 = k_1, \cdots, X_n = k_n)}{\Pr(X_i = 1)\cdot\Pr(X_1 = k_1, \cdots, X_n = k_n)} = \\ &\frac{\Pr(T = 1, X_i = 1, X_1 = k_1, \cdots, X_n = k_n)}{\Pr(X_i = 1)}\cdot 2^{n-1}\end{aligned} \tag{12}$$

Similarly, the last half of the Eq. (11) can be expressed as:

$$\begin{aligned}&\Pr(T = 1 \mid X_i = 0, X_1 = k_1, \cdots, X_n = k_n) = \\ &\frac{\Pr(T = 1, X_i = 0, X_1 = k_1, \cdots, X_n = k_n)}{\Pr(X_i = 0)}\cdot 2^{n-1}\end{aligned} \tag{13}$$

Then put the Eq. (12) and the Eq. (13) into the Eq. (11), the result is the same as Eq. (7).

$$\begin{aligned}I_{\phi(i)} &= \frac{1}{2^{n-1}} \sum_{k_1,k_2,\cdots,k_{i-1},k_{i+1},\cdots,k_n=0}^{1} \left[\frac{\Pr(T = 1, X_i = 1, X_1 = k_1, \cdots, X_n = k_n)}{\Pr(X_i = 1)}\cdot 2^{n-1} \right. \\ &\left. - \frac{\Pr(T = 1, X_i = 0, X_1 = k_1, \cdots, X_n = k_n)}{\Pr(X_i = 0)}\cdot 2^{n-1}\right] \\ &= \sum_{k_1,k_2,\cdots,k_{i-1},k_{i+1},\cdots,k_n=0}^{1} \left[\frac{\Pr(T = 1, X_i = 1, X_1 = k_1, \cdots, X_n = k_n)}{\Pr(X_i = 1)} \right. \\ &\left. - \frac{\Pr(T = 1, X_i = 0, X_1 = k_1, \cdots, X_n = k_n)}{\Pr(X_i = 0)}\right] \\ &= \frac{\sum_{k_1,k_2,\cdots,k_{i-1},k_{i+1},\cdots,k_n=0}^{1} \Pr(T = 1, X_i = 1, X_1 = k_1, \cdots, X_n = k_n)}{\Pr(X_i = 1)} \\ &- \frac{\sum_{k_1,k_2,\cdots,k_{i-1},k_{i+1},\cdots,k_n=0}^{1} \Pr(T = 1, X_i = 0, X_1 = k_1, \cdots, X_n = k_n)}{\Pr(X_i = 0)} \\ &= \frac{\Pr(T = 1, X_i = 1)}{\Pr(X_i = 1)} - \frac{\Pr(T = 1, X_i = 0)}{\Pr(X_i = 0)} \\ &= \Pr(T = 1 \mid X_i = 1) - \Pr(T = 1 \mid X_i = 0)\end{aligned}$$

3.2. The Probability Importance Degree

In a $\overline{FT}$, the probability importance degree of bottom-event x_i can be calculated by the Eq. (14) [14].

$$I_g(i) = \frac{\partial g(q)}{\partial q_i} \tag{14}$$

Here q_i represents the occurrence probability of the *ith* bottom-event.

In an equivalent $\overline{BN}$, the probability importance degree of the point X_i can be calculated by the Eq. (15) [7].

$$I_g(i) = \Pr(T=1 \mid X_i = 1) - \Pr(T=1 \mid X_i = 0) \tag{15}$$

Furthermore, for any BN, if its points' values are all binary, this BN can be expressed as a multivariate linear equation (MLE) [8,16,17], which is a foundation of the Eq. (15). At the same time, each MLE contains two types of parameters: evidence indicators and network parameters. The meaning of the evidence indicator is that in a BN, any variable X corresponds to an evidence indicator λ_x. The meaning of the network parameter is that in a BN, any variable X's family group XU corresponds to a parameters set $\theta_{x|u}$, which is the probability parameter in the CPT.

Definition 2 gives an accurate definition of the MLE in a BN as follow [8].

Definition 2: for any BN, if its variable set is W; and for any variable X's family group XU, if its set is W as well. The BN's corresponding MLE can be defined as follow.

$$f \overset{def}{=} \sum_{z} \prod_{\theta_{x|\boldsymbol{u}} : z} \theta_{x|\boldsymbol{u}} \prod_{x : z} \lambda_x$$

In definition 2, $\theta_{x|u}$: z represents the values of $\boldsymbol{xu}$ and z are compatible, x : z represents the values of x and z are compatible. So, according to definition 2, the BN in Fig.1 can be expressed as follow.

$$f = \lambda_a\lambda_b\theta_a\theta_{b|a} + \lambda_a\lambda_{\overline{b}}\theta_a\theta_{\overline{b}|a} + \lambda_{\overline{a}}\lambda_b\theta_{\overline{a}}\theta_{b|\overline{a}} + \lambda_{\overline{a}}\lambda_{\overline{b}}\theta_{\overline{a}}\theta_{\overline{b}|\overline{a}}$$

Definition 3: for each BN, given the evidence variable e, set the value of evidence indicator λ_x as 1, if it is compatible with e; set the value of evidence indicator λ_x as 0, if it is not compatible with e. Then, the following equations can be obtained [8]:

$$f \overset{def}{=} f(\boldsymbol{e}) \text{ and } f(\boldsymbol{e}) = \Pr(\boldsymbol{e})$$

As the same example in Fig.1, if the evidence variable e=a, set $\lambda_a = 1, \lambda_{\overline{a}} = 0, \lambda_b = 1, \lambda_{\overline{b}} = 1$ in MLE, then the $f(a) = \theta_a\theta_{b|a} + \theta_a\theta_{\overline{b}|a} = \Pr(a)$ can be obtained.

So a proof for the Eq. (15) is as follows based on the knowledge as described above:

The Eq. (16) is correct according to the MLE as described above. Then the Eq. (17) can be obtained by calculating the partial derivative for the parameter $\theta_{y|v}$.

$$\Pr(\boldsymbol{e}) = f(\boldsymbol{e}) = \sum_{z:\,e} \prod_{\theta_{x|u} : z} \theta_{x|\boldsymbol{u}} \tag{16}$$

$$\frac{\partial f(\boldsymbol{e})}{\partial \theta_{y|v}} = \sum_{\substack{z:\, yv \\ z:\, e}} \prod_{\substack{\theta_{x|u} : z \\ xu \neq yv}} \theta_{x|\boldsymbol{u}} - \sum_{\substack{z:\, \overline{y}v \\ z:\, e}} \prod_{\substack{\theta_{x|u} : z \\ xu \neq \overline{y}v}} \theta_{x|\boldsymbol{u}} \tag{17}$$

Moreover, two sub-equations in right side of the Eq. (17) can be translated to the Eq. (18) and the Eq. (19), respectively.

$$\sum_{\substack{z:\, yv \\ z:\, e}} \prod_{\substack{\theta_{x|u} : z \\ xu \neq yv}} \theta_{x|\boldsymbol{u}} = \sum_{\substack{z:\, yv \\ z:\, e}} \Big(\prod_{\theta_{x|u} : z} \theta_{x|u} / \theta_{y|v} \Big) = \sum_{\substack{z:\, yv \\ z:\, e}} \frac{\Pr(z)}{\theta_{y|v}} \tag{18}$$

$$\sum_{\substack{z:\, \overline{y}v \\ z:\, e}} \prod_{\substack{\theta_{x|u} : z \\ xu \neq \overline{y}v}} \theta_{x|\boldsymbol{u}} = \sum_{\substack{z:\, \overline{y}v \\ z:\, e}} \Big(\prod_{\theta_{x|u} : z} \theta_{x|u} / \theta_{\overline{y}|v} \Big) = \sum_{\substack{z:\, \overline{y}v \\ z:\, e}} \frac{\Pr(z)}{\theta_{\overline{y}|v}} \tag{19}$$

Then put the Eq. (18) and the Eq. (19) into the Eq. (17) to get the Eq. (20).

$$\frac{\partial f(\boldsymbol{e})}{\partial \theta_{y|v}} = \sum_{\substack{z:\, yv \\ z:\, e}} \frac{\Pr(z)}{\theta_{y|v}} - \sum_{\substack{z:\, \overline{y}v \\ z:\, e}} \frac{\Pr(z)}{\theta_{\overline{y}|v}} = \frac{\Pr(y,\boldsymbol{v},\boldsymbol{e})}{\theta_{y|v}} - \frac{\Pr(\overline{y},\boldsymbol{v},\boldsymbol{e})}{\theta_{\overline{y}|v}} \tag{20}$$

Furthermore, in a $\overline{FT}$, the equation for probability importance degree is $I_g(i)=\partial g(q)/\partial q_i$; in an equivalent $\overline{BN}$, the $g(q)=\Pr(T=1)$ and the $q_i=\Pr(X_i=1)$; in a MLE, the $\Pr(X_i=1)=\theta_{x_i}$, the $\Pr(X_i=0)=\theta_{\overline{x}_i}$ and the $f(\boldsymbol{e})=f(T=1)=\Pr(T=1)$. So the Eq. (21) can be got as follow.

$$I_g(i)=\frac{\partial g(q)}{\partial q_i}=\frac{\partial \Pr(T=1)}{\partial \Pr(X_i=1)}=\frac{\partial f(\boldsymbol{e})}{\partial \theta_{x_i}} \quad (21)$$

According to the Eq. (20), the Eq. (21) can be translated as follow, which is the same as the Eq. (15).

$$\begin{aligned} I_g(i)&=\frac{\partial f(\boldsymbol{e})}{\partial \theta_{x_i}}=\frac{\Pr(x_i,\boldsymbol{e})}{\theta_{x_i}}-\frac{\Pr(\overline{x}_i,\boldsymbol{e})}{\theta_{\overline{x}_i}}=\frac{\Pr(x_i,\boldsymbol{e})}{\Pr(x_i)}-\frac{\Pr(\overline{x}_i,\boldsymbol{e})}{\Pr(\overline{x}_i)} \\ &=\Pr(\boldsymbol{e}\,|\,x_i)-\Pr(\boldsymbol{e}\,|\,\overline{x}_i) \\ &=\Pr(T=1\,|\,X_i=1)-\Pr(T=1\,|\,X_i=0) \end{aligned} \quad (22)$$

3.3. The Probability Importance Degree

In a $\overline{FT}$, the critical importance degree of the bottom-event x_i can be calculated by the Eq. (23) [14].

$$I_c(i)=I_g(i)\frac{q_i}{g(q)} \quad (23)$$

While an equivalent $\overline{BN}$, the critical importance degree of the X_i can be calculated by the Eq. (24) [7].

$$I_c(i)=\frac{\Pr(X_i=1)[\Pr(T=1\,|\,X_i=1)-\Pr(T=1\,|\,X_i=0)]}{\Pr(T=1)} \quad (24)$$

A proof for the Eq. (24) is as follow.
Put the Eq. (22) into the Eq. (23) to get the equation which is the same as the Eq. (24).

$$\begin{aligned} I_c(i)&=[\Pr(T=1\,|\,X_i=1)-\Pr(T=1\,|\,X_i=0)]\text{g}\frac{q_i}{g(q)} \\ &=[\Pr(T=1\,|\,X_i=1)-\Pr(T=1\,|\,X_i=0)]\text{g}\frac{\Pr(X_i=1)}{\Pr(T=1)} \end{aligned}$$

4. Further study

The Eq. (5), the Eq. (7), the Eq. (15) and the Eq. (24) are all belong to the prior/posterior probability issues of the BN. For solving the posterior probability issue, the prior probability has to be solved firstly (see the Eq. (3)). At the same time, the solutions of the prior/posterior probability issues are all belong to the solution of the marginal probability $\Pr(\boldsymbol{Y}=\boldsymbol{y})$, where $\boldsymbol{Y}$ represents some points' sets for a BN. So, a unified algorithm can be used by applying the BN method to analyse the FT, which is better than the normal FT methods. Because in the normal FT methods, different problems have to be solved by different methods rather than the unified algorithm.

Moreover, in addition to calculating three importance degrees and the failure probability of the top-event, the BN method can also calculate the minimal cut/path sets to analyse the FT [7]. Furthermore, the BN method can be used for causal reasoning (see the Eq. (3)) and fault diagnosis (see the Eq. (4)) due to the characteristics of the Bi-directional reasoning. At the same time, the Eq. (3) and the Eq. (4) can be transferred to the Eq. (25) and the Eq. (26), respectively, and $\boldsymbol{X}=\boldsymbol{x}$ represents arbitrary point set and its value in a BN. It means that the BN method not only can calculate the normal working probability of a network by given any one or more points, but also can calculate the failure probability of any one or more points to judge the weakness of the network when it is fault. That is, more information about the whole system could be acquired by the BN method, which is better than the FT method.

$$\Pr(T=1\,|\,\boldsymbol{X}=\boldsymbol{x})=\frac{\Pr(T=1,\boldsymbol{X}=\boldsymbol{x})}{\Pr(\boldsymbol{X}=\boldsymbol{x})} \quad (25)$$

$$\Pr(\boldsymbol{X}=\boldsymbol{x}\,|\,T=1)=\frac{\Pr(T=1,\boldsymbol{X}=\boldsymbol{x})}{\Pr(T=1)} \quad (26)$$

Finally, the FT method is only suitable for a system which has a determined fault mechanism and a clear causal logic. Moreover, the FT method is based on three hypotheses [6,18]: only two states (0/1) for an event; the determined causal logic between events (one event causes another event definitely or not); the fault independence of the bottom-event. While the application of the BN method can transform the FT into the BN to modify the network further to get rid of these three hypotheses as described above [18].At the same time, the BN method can also reconstruct a new network directly according to the causal logic of points in the network to free from these three hypotheses as well [12].

5. Case study

The Fig. 2(a) is a FT of an aircraft landing gear [14] and the Fig. 2(b) is an equivalent BN.

5.1. Results

The unreliability (the occurrence probability of the top-event) of system is $\Pr(T=1)=1.25047\times10^{-5}$ by the Eq. (5). The structure importance degree, the probability importance degree, the critical importance degree, the causal reasoning and the diagnostic reasoning are calculated by the Eq. (2), the Eq. (15) and the Eq. (24), the Eq. (3) and the Eq. (4), respectively. All results are shown in Table I.

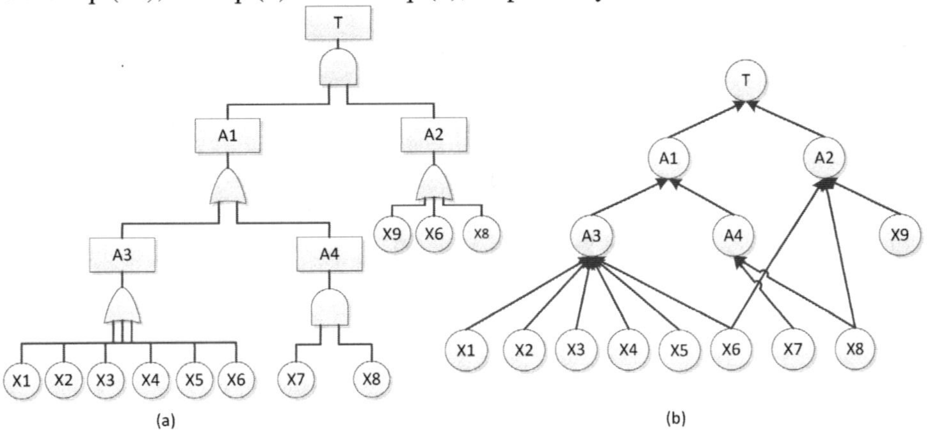

Figure 2. (a) The FT of an aircraft landing gear and (b) The equivalent BN.

Table 1. The arrangement of channels

B.	S.	P.	Cr.	Ca.	D.
X1	0.0156	6.3197e-5	1.2635e-5	6.3199e-5	1.2634e-4
X2	0.0156	6.3198e-5	6.3174e-5	6.3199e-5	6.3171e-5
X3	0.0156	6.3199e-5	5.0540e-13	6.3199e-5	5.0537e-13
X4	0.0156	6.3194e-5	3.6134e-4	6.3199e-5	3.6134e-4
X5	0.0156	6.3199e-5	1.0108e-12	6.3199e-5	1.0107e-12
X6	0.0156	0.9999	0.9996	0.9999	0.9996
X7	0.0078	2.7504e-5	4.3989e-8	2.7504e-5	4.3987e-8
X8	0.2500	1.2150e-4	1.4576e-4	1.2152e-4	1.4575e-4
X9	0.2422	1.2149e-4	3.4685e-4	1.2150e-4	3.4684e-4

B.=Bottom-event, S.=Structure importance degree, P.=Probability importance degree, Cr.=Critical importance degree, Ca.=Causal reasoning and D.=Diagnostic reasoning.

5.2. Discussion

- In order to avoid the complex disjoint method, Deng used a direct calculation method to get the occurrence probability of the top-event approximately [14]. In contrast, the BN method can acquire more accurate results, which is simple and easy.
- Table I shows more accurate results of the structure importance degree, while in Deng's paper [14], the structure importance degree was roughly obtained by the minimal cut/path sets. That

is why two results are slightly different. Compared with the traditional FT method, the BN method can get more accurate results, and the calculation of the structure importance degree is much simpler.

- For the complex FT, the probability function is difficult to obtain, which not only leads to the inaccurate reliability, but also cause the inaccurate probability importance degree and critical importance degree of the bottom-event. While there is no such shortage for the BN method, and more accurate results of the probability importance degree and critical importance degree can be obtained through the Eq. (15) and the Eq. (24).
- The BN method not only can obtain the results that calculated by the traditional FT method, but also can use the causal reasoning and diagnostic reasoning to analyse the reliability of the system. Because the causal reasoning can be used to determine the influence for the system caused by the bottom-event, and the diagnostic reasoning can diagnose the system to identify the weak points (for a system, the higher probability of $Pr(X=1|T=1)$, the weaker point it is).

6. Conclusion

Combined with the characteristics of the BN method, this paper systematically study the reasoning processes for calculating the structure importance degree, the probability importance degree and the critical importance degree. A case study has demonstrated that the BN method is better than the conventional FT method. The conclusions for the BN method are as follows.

- The BN method can use a unified algorithm to get richer information and more accurate results, which is important for the fault diagnosis and the reliability assessment.
- The BN method overcomes the shortcomings of the FT method since it does not rely on the three hypotheses. That is, the BN method is more comprehensive and accurate for modelling complex systems, and it also can use an original reasoning method to analyse and process complex systems as well.

7. References

[1] Wang H, Zhou J L and He Z Y 2004 *J. Compt. Integr. Manuf. Syst.* **10(2)** 230-234
[2] Yi X W, Qian W X and Xie L Y 2008 *J. Acta Aeronautica et Astronautica Sinica* **29(6)** 1482-1489
[3] Torres-toledano J G and Sucar L E 1998 in *Proc. 1998 IBERAMIA Conf.* p 195-206
[4] Khakzad N, Khan F and Amyotte P 2011 *J. Reliab. Eng. Syst. Saf.* **96(8)** 925-932
[5] Mahadevan S, Zhang R and Smith N 2001 *J. Struct. Saf.* **23(3)** 231-251
[6] Portinale L and Bobbio A 1999 in *Proc. 1999 UAI Conf.* p 551-558
[7] Zhou Z B, Dong D D and Zhou J L 2006 *J. Syst. Eng. Theory. Pract.* **26(6)** 95-100
[8] Darwiche A 2012 *Modeling and Reasoning with Bayesian Networks* (New York: Cambridge University Press) p 53-306
[9] Shi Z F and Zhang A 2012 *Theory of Bayesian network and its application in military system* (Beijing: National Defense Industry Press) p 78-104
[10] Jensen F V 1996 *An Introduction to Bayesian Networks* (New York: Springer-Verlag) p 1-178
[11] Bobbio A, Portinale L and Minichino M 2001 *J. Reliab. Eng. Syst. Saf.* **71(3)** 249-260
[12] Mengshoel O J, Chavira M and Cascio K 2010 *J. IEEE Trans Syst. Man Cybern. A* **40(5)** 874-885
[13] Huang C and Darwiche A 1996 *J. Int. J. Approx. Reason.* **15(3)** 225-263
[14] Deng Q 2009 *Safety system engineering* (Xi'an: Northwest Industrial University press) p 95-97
[15] Song B F 2006 *Reliability engineering of aircraft* (Xi'an: Northwest Industrial University press) p 77-93
[16] Chavira M and Darwiche A 2007 in *Proc. 2007 IJCAI Conf.* p 2443-2449
[17] Chavira M and Darwiche A 2005 in *Proc. 2005 IJCAI Conf.* p 1306-1312
[18] Langseth H and Portinale L 2007 *J. Reliab. Eng. Syst. Saf.* **92(1)** 92-108

IOP Conf. Series: Materials Science and Engineering **362** (2018) 012026 doi:10.1088/1757-899X/362/1/012026

Utilization of Additive Manufacturing in Evaluating the Performance of Internally Defected Materials

A-H I Mourad[1*], A M Ghazal[2], M M Syam[2] and O D Al Qadi[2] and H Al Jassmi[3]

1 Professor, Department of Mechanical Engineering, UAE University, PO. Box 15551 Al-Ain, UAE
2 Research Scholar, Department of Mechanical Engineering, UAE University, PO. Box 15551 Al-Ain, UAE
3 Assistant Professor, Department of Civil & Environmental Engineering, UAE University, PO. Box 15551 Al-Ain, UAE
* Corresponding author e-mail: ahmourad@uaeu.ac.ae

Abstract. The elimination of internal defects in a material present in the raw material or generated during the manufacturing or service is difficult. The inclusions of the defects have an adverse effect on the load bearing capacity. The presence of the cracks subjected to a specific orientation in materials or machinery can cause devastating unexpected failure during operation. Analysis of the failure in the components with cracks is more confined to analytical and numerical evaluation. The experimental evaluation has been tedious due to the complexity of replicating the actual defected component. The potential of additive manufacturing in developing user-defined components with cracks for the experimental evaluation is explored in this research. The present research investigated the effect of the internal elliptical cracks aligned at different orientations on the mechanical performance of polylactic acid (Green filament). The Fusion Deposition Method was utilized for the development of the standard tensile specimens with internal elliptical crack oriented at 0°, 45° and 90° using UltiMaker 2. The results proved that there is a considerable reduction in the load bearing capacity due to the presence of the cracks. The maximum load bearing capacity decreased by 15.01% for the specimen with crack inclined at 0° to the lateral axis compared to crack- free specimen. The nature of the fracture and the stress-strain graph evidently showcase the brittle nature of the material. The SEM image of the fractured region proved the phenomenal characteristics such as strong adhesion between the layers and the proper material flow. In the light of the results of this work, it can be concluded that the 3-D printing methodology is effective for evaluating the mechanical performance of the internally defected material.
Keywords: Additive Manufacturing (AM); Fused deposition method (FDM); Infill rate; Internal elliptical crack; Mechanical performance.

1. Introduction

There is a crucial need for examining the mechanical properties of structures with internal defects especially in applications were such tiny defects can cause catastrophic failure (i.e., Aerospace or biomedical applications). The above-mentioned problem was solved analytically in the absence of conventional fabrication methods to produce samples with internal defects for experimental testing. For more than a decade, additive manufacturing (AM) or 3D printing has taken the attention due to its beneficial ability to handle a high level of complexity in terms of geometry and material, which could

IOP Conf. Series: Materials Science and Engineering **362** (2018) 012026 doi:10.1088/1757-899X/362/1/012026

not be fabricated by subtractive manufacturing methods in rapid prototyping (RP). This technology is defined as a process for producing three-dimensional (3D) objects from Computer-Aided Design (CAD models) through an additive process in which successive layers adds sequentially [1]. Fused deposition modelling (FDM), selective laser melting (SLM) and selective laser sintering (SLS) are commonly used techniques to take advantage of material extrusion. Additional to variant methods of 3D printing technologies in the early 1990s, the Laminated Object Manufacturing (LOM) has developed by Helisys of Torrance which provides a laminated concept in 3D printing [2].
Among AM technologies, the fused deposition modelling (FDM) has the largest share from the most widely used techniques. This technique has been firstly developed by Stratasys Inc. since last century. According to the 2011 Wohler's Report, the FDM currently has reached 41.5 percent of the used 3D systems commercially [3]. The technique used the advantage of thermoplastic materials by continuous extrusion and deposition with solidification properties of the material [4]. FDM has three key elements in the system including material feed mechanism, liquefier/print head and building surface.
Figure 1 shows the main components of FDM 3D printer [5, 6]. Two materials are commonly processed by such technology namely acrylonitrile butadiene styrene (ABS) and polylactic acid (PLA) as FDM filaments, which will be extruded through the liquefier to the printing nozzle which is illustrated in flow diagram depicted in figure 2 [7]. PLA is proven to be a stronger alternative than ABS for FDM technologies as reported in the literature. H. Kim et al. [7] reported that PLA has higher tensile strength as compared to ABS. It was noted that the liquefied PLA thread has a stronger bonding than the liquefied ABS thread, which was observed from the failure surface.

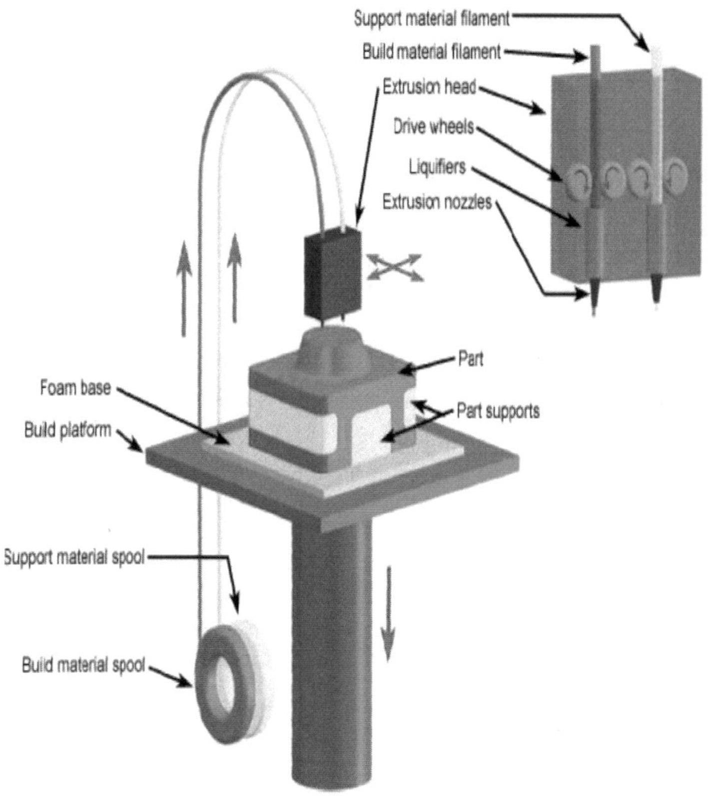

Figure 1. Fused deposition modeling 3D printer [6]

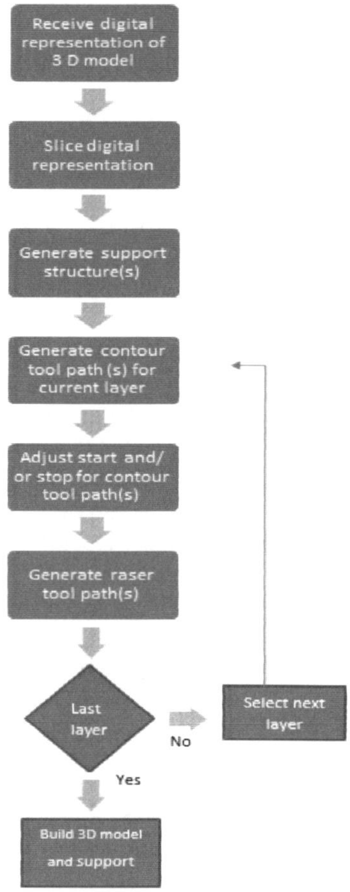

Figure 2. FDM procedure diagram

Polylactic acid (PLA) is a biodegradable thermoplastic derived from sugar and starch. It has malleable behaviour when it is heated, which allows for various shapes' creation. As far as the mechanical properties are concerned, PLA is a brittle material that has relatively high tensile and flexural moduli and has good stiffness and strength [8]. Letcher and M. Waytashek [9] have tested 3D printed PLA material of different raster angles (0°, 45° and 90°) under tension, flexural and fatigue loadings to evaluate the material's properties. Tensile tests revealed that the 45° raster angle resulted in the strongest fabricated samples. However, the flexural testing showed that the raster angle of 0° resulted in the highest bending stress among the tested samples. As far as the fatigue testing results are concerned, both 0° and 45° raster angles resulted in almost similar behaviour that was better than the one attained at 90°. In addition, B. Witt Brodt and J. Pearce [10] have investigated the effect of the PLA colour (i.e., natural, white, blue, grey and black) on the mechanical properties of the material. Conducted tensile testing (refer to table 1) showed that the highest strength occurred for natural colour, whereas the lowest one was attained by the grey colour. It should be noted that the authors had assessed the effect of extrusion temperature (i.e., 190 ^oC to 215 ^oC) on the tensile strength and crystallinity percentage for the white colour. The results had revealed that the increase in extrusion temperature results in increasing the tensile strength in general; however, the 200 ^oC tests showed out-of-trend behaviour, which was justified by the preference direction of the developed grains. Also, there was an existence for a maximum extrusion temperature at which the maximum crystallinity percentage occurs.

Table 1. Properties of PLA [10].

Colour	Ultimate tensile strength (MPa)	Yield strength (MPa)	Maximum strain (%)	Crystallinity (%)
Natural	57.16 ± 0.35	52.47 ± 0.35	2.35 ± 0.05	0.93 ± 0.06
Black	52.81 ± 1.18	49.23 ± 1.18	2.02 ± 0.08	2.62 ± 0.09
Grey	50.84 ± 0.23	46.08 ± 0.23	1.98 ± 0.04	4.79 ± 0.10
Blue	54.11 ± 0.30	50.10 ± 0.30	2.13 ± 0.02	4.85 ± 0.15
White	53.97 ± 0.26	50.51 ± 0.26	2.22 ± 0.04	5.05 ± 0.18

J. Chacón et al. [11] have investigated the effect of 3D printing process parameters on the mechanical properties of printed PLA structures. In this study, building orientation, layer thickness and feed rate were varied and their consequence effect on the tensile and three-point bending was examined. D. Farbman and C. McCoy [12] have studied the effect of infill percentage (0% up to 100%), infill orientation (rectilinear and hexagonal) and strain rate (up to 15 mm/second) on the mechanical performance of ABS and PLA material. It was concluded that the higher infill percentages result in stronger samples.

The main objectives of the research are to produce 3D printed standard tensile samples with and without internal elliptical defects and to examine the effect of elliptical internal defects on the mechanical performance of the 3D printed PLA structures considering different orientations (0°, 45° and 90°). The capability of utilizing additive manufacturing for evaluating the mechanical performance of defected materials could be a major advantage in near future. The production of defect-free specimens with high quality are mandatory for adopting the technique. The effect of the parameters on the quality is dominant and the current study evaluates the afore-mentioned aspects.

2. Materials and Methods

In the present work, a commercial type of green PLA filament with 2.85mm diameter was opted for the experimentations. The mechanical properties of the PLA manufactured by FDM is shown in table 2.

Table 2. Mechanical properties of PLA [13, 14].

Properties	Value
Tensile strength (MPa)	15.5-72.2
Tensile modulus (GPa)	2.020-3.550
Elongation at break (%)	0.5-9.2

ASTM D638 standard for the polymers testing was followed due to lack of an international standard for the evaluation of the mechanical properties of 3D printed specimens. The tensile specimen model with internal elliptical cracks was modelled in CATIA V5 and exported to .stl format. Figure 3 shows the geometrical dimensions of the standard tensile specimen.

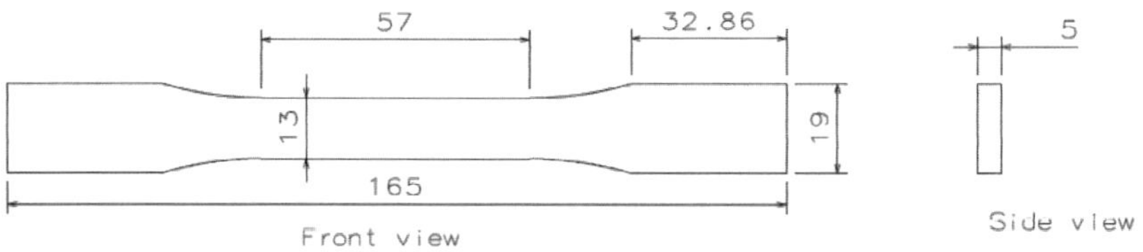

Figure 3. Geometrical dimension of the tensile specimen. All dimensions in mm.

Figure 4 presents a 3D schematic drawing showing the crack orientations. The crack length (2c) and crack width (2a) are 2mm and 1mm, respectively, for the internal elliptical crack.

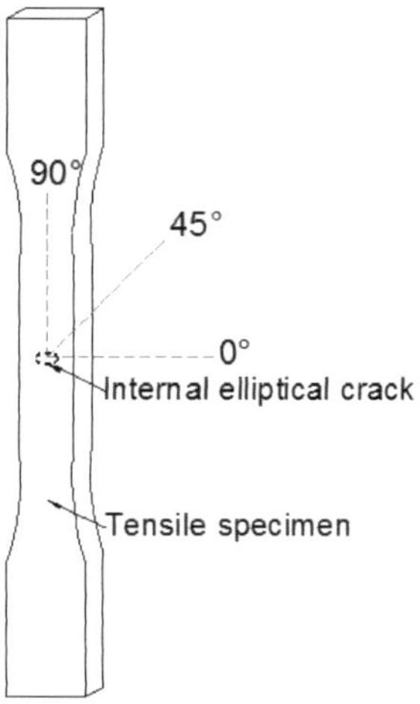

Figure 4. 3D Schematic drawing of the tensile specimen showing different crack orientations.

Different specimens with crack orientations of 0°, 45° and 90° to the lateral axis are fabricated. The results are compared with the crack-free specimens.

2.1 Process Parameters

PLA colour type, raster angle, printing layer thickness, feed rate and infill rate are the major process parameters used for FDM 3D printing process. These used process parameters are given in table 3.

Table 3. Process parameter for FDM technique.

PARAMETERS	VALUE / SPECIFICATION
PLA (colours)	Green
Layer thickness (mm)	0.1
Shell wall thickness (mm)	1.05
Shell top/bottom thickness (mm)	0.8
Infill percentage (%)	100 %
Feed rate/printing speed (mm/s)	80
Raster angle	45°
Extruding temperature	228°C

2.2 Tensile Testing

The PLA specimens were tested using E45.305 tensile testing machine produced by MTS and equipped with 100 KN load cell. The tests were conducted at 0.25 mm/s. Total of eighteen tensile samples was tested under standard room temperature (25°C) since the thermoplastic material is very dependent on the temperature. In addition, MTS axial extensometer (MODEL 632,25F-20) has been used for strain measurements.

2.3 Scanning Electron Microscopy (SEM)

Scanning Electron Microscopy (SEM) images were taken at the fractured surfaces after the tensile testing. The SEM - JSM-6010LA developed by JEOL was used in this study. Generally, the SEM images are capable of providing high resolution with fine detailed images to identify and quantify the compositional information and surface topography. The machine used in this study can provide a resolution of 4 nm and a magnification power ranging from 5X up to 300,000X. The device can handle a specimen size up to 150 mm diameter [15].

3. Results and Discussions

The presence of the internal cracks is expected to lower down the load bearing capacity of the material. In the presence of the crack, it is more vital to evaluate the maximum load bearing capacity of the material before fracture than the tensile strength of the specimen. The tensile strength is analytically calculated as the ratio of peak tensile force to the cross-sectional area. The load-displacement curve and stress-strain graph for the crack-free specimen are shown in figure 5. The lack of sufficient yielding before fracture in the stress-strain graph proves the brittleness of the PLA-green specimen. The modulus of elasticity for the non-defected PLA-green sample calculated from the stress-strain graph is 2774 MPa. This value is within the ranges reported in the table 2 (2020 MPa – 3550 Mpa).

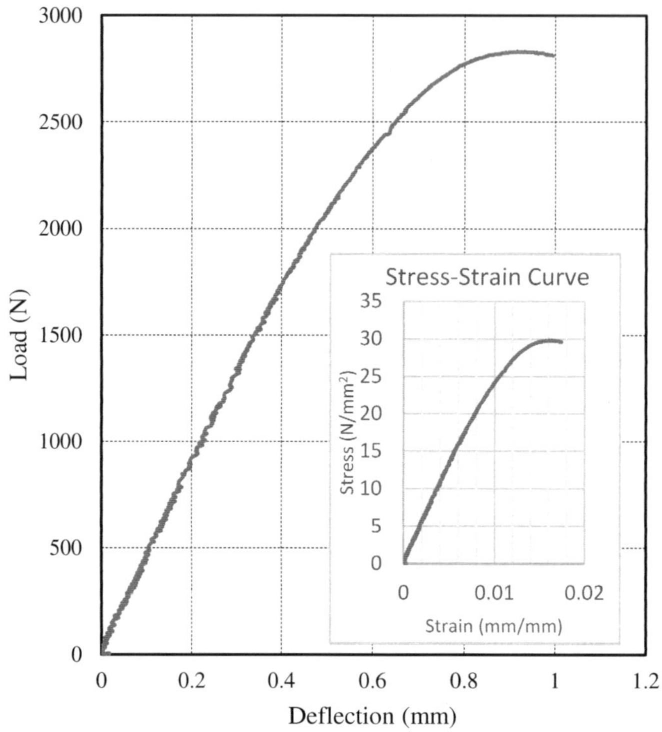

Figure 5. Load-deflection curve and stress-strain graph for crack-free PLA specimen (Green)

The presence of the cracks lessens the cross-sectional area of the sample. The determination of the tensile strength is inappropriate in such circumstances. In such situation, two approaches can be used to evaluate the load bearing capacity of the specimen. These are the fracture mechanics approach in which the stress intensity factor is the characterizing parameter or the stress concentration approach. Figure 6 shows the load-displacement curves obtained for the PLA-green specimens with crack oriented at 0°, 45° and 90° and without a crack. The graphs clearly indicated the brittle nature of the specimen. The specimens with crack and without crack resulted in the same trend. Three separate regions may be classified based on the load displacement distributions. The stiffness (Young's modulus) is constant in the first region in which the Hooke's law is applicable whereas the stiffness decreases slightly in the second region. There is more reduction in the stiffness in the third region before the fracture. The load-displacement curves indicate that the crack free specimen possess more load withstand ability. The cracks found to depreciate the load carrying capacity of the material. The result signifies the relevance of avoiding the cracks from the materials that are subjected to load conditions. The orientation has found to have profound relevance in terms of the load bearing capacity.

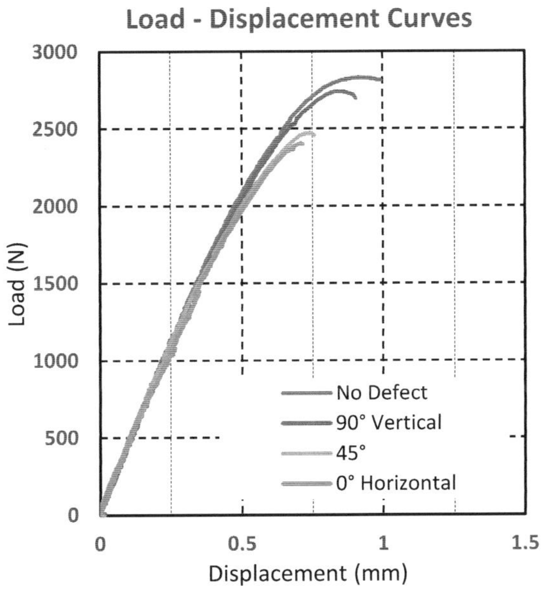

Figure 6. Force - Displacement of the PLA (green)

The load bearing capacity of the sample with crack oriented 90° to the lateral axis is more than the specimen having the crack oriented 0° to the lateral axis. The load bearing capacity of the PLA-green specimen with crack inclined 45° to the lateral axis is nearly close to that of the specimen with crack oriented at 0°.

The bar chart shown in figure 7 shows the maximum load carrying capacity of the crack and crack-free specimens. The maximum withstanding load before the fracture for the defect free specimen was 2830 N whereas the peak loads were 2743 N, 2472 N and 2405 N for the specimens with crack inclined at 90°, 45° and 0° respectively. The peak load decreased as the angle of orientation to the lateral axis is decreased. A reduction of 15.01% in the peak load, compared to the defect-free sample, is recorded for the sample of 0° orientation. These peak load values are considerably reduced compared to cracked specimen oriented 90° to the lateral axis. The closeness of the peak load for the defect-free and 90° crack-oriented specimen is also observed. The load-displacement graph reveals that the defect size was found to have no influence on the slope of the load-displacement graph (i.e., stiffness of the material) .

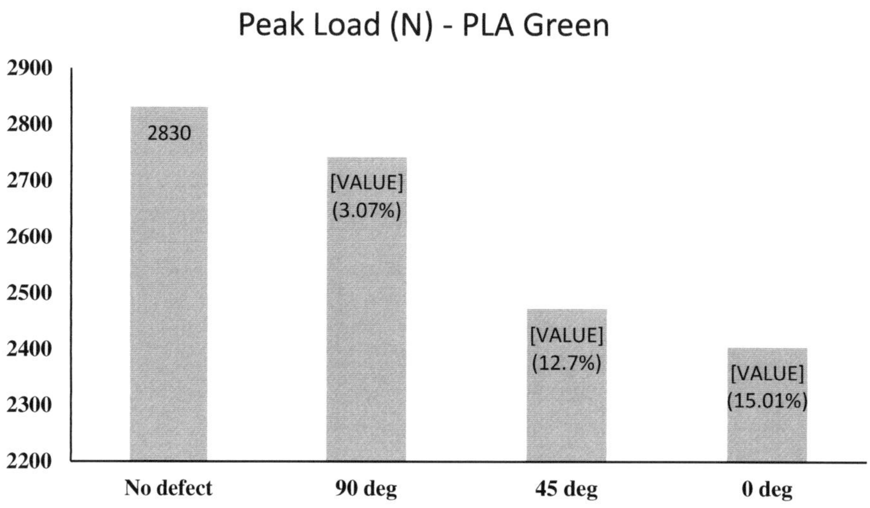

Figure 7. Effect of the defect's orientation on the peak load

The reduction of the peak load when the crack inclines towards the lateral axis causes an increase in the normal opening stress which is the weak link of brittle material. The internal defected samples were fractured exactly at the center of the gauge length where the crack was present as shown in figure 8. It is observed that the mode of fracture is brittle and there is no evidence for any necking which support the observation from the load-displacement curves.

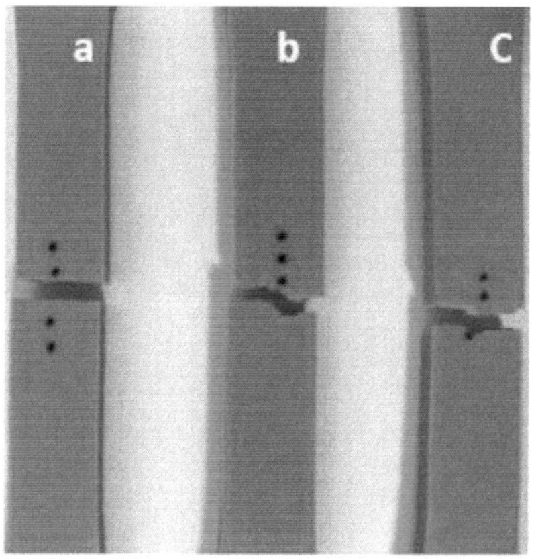

Figure 8. Fractured samples of samples oriented at (a) 0°, (b) 45°, and (c) 90°

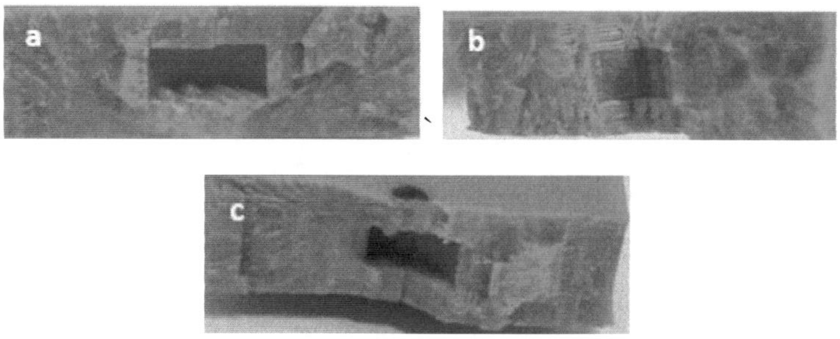

Figure 9. Fracture surfaces (a) 0°, (b) 90°, and (c) 45°

Figure 9 shows the fractured surfaces of the specimen with cracks oriented at different angles. The fracture surface topography shows that the material is porous-free with smooth appearance. This is the main reason of getting the expected trend and the high level of repeatability of the results. Such evidence prove that the 3D printing technique can be used to produce quality samples that can be employed for studying and evaluating the mechanical properties of defected and defect-free materials. The enlarged optical image and the SEM image of the fractured surface are shown in figure 10 and figure 11 respectively. Figure 10 shows that the fracture surface is smooth and porous-free to a large extent. This reflects the proper flow of the molten material.

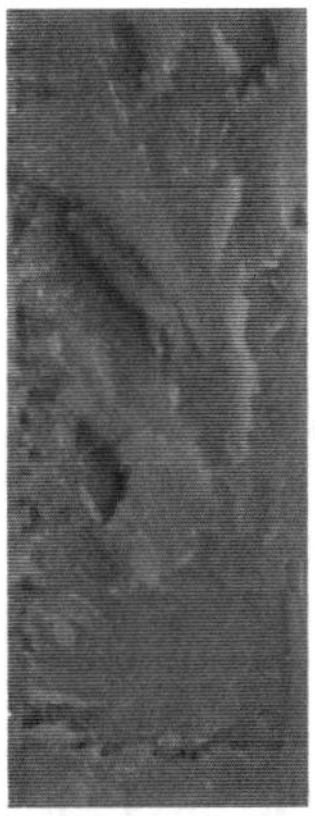

Figure 10. Magnified Optical Images (10x) at the fractured surface – PLA Green

Figure 11. SEM Image of the fractured surface - PLA Green (25x)

The melting and solidification of the material that has been laid by layers is the fundamental characteristic of additive manufacturing. The SEM image of the fracture surface in figure 11 shows strong adhesion between the layers and the proper flow of the material. This could result in the absence of the layering and the pores that may be suspected even at 100% infill rate. Presence of the voids or gaps anticipated at lower infill rate could deteriorate the mechanical performance. There is a smooth build-up of the material far from the edges or shell material. The smoother builds up imply stronger inter-layer fusion bonds between the different layers; hence, expected to have higher stress values [16]. No layer separation showcases the unique property of the 3D printing to develop materials layer by layer forming a defect-free solid rigid object that can be used to assess the mechanical characterization of the material with and without defects. The results shows also that the additive manufacturing/3D printing method can be utilized to characterize the stable crack growth through materials [17-21] with minimum number of samples.

4. Conclusion

The research focused on examining the effect of internal elliptical crack on the mechanical performance of 3D-Printed PLA (Green filament). Crack and crack-free specimens fabricated according to ASTM D638 were printed using UltiMaker 2. The specimens with elliptical crack inclined 0°, 30° and 90° to the lateral axis are investigated. The crack-free specimen has the maximum load capacity. The load bearing capacity reduced considerably due to the presence of internal elliptical cracks. The reduction of the maximum load for the specimen with cracks oriented at 0°, 45° and 90° to the lateral axis compared to crack-free PLA (green filament) specimen accounted to 15.07%, 12.7% and 3.07% respectively. The defected samples fractured exactly at the center of the gauge length where the crack was present. The SEM images of the fractured surfaces are captured. No layering in the SEM image of the fractured surface showcase strong adhesion among the layers. The smooth buildup of the material from the edges or shell is noticed. The results of this work proved that the additive manufacturing is a promising technique for creating well specified internal cracks in a specimen or component. Therefore, it can be utilized in accurately evaluating the mechanical performance of materials. The number of experiment runs can be reduced for the repeatability tests. This will reflect on the cost of the test. Less number of experiment runs, low cost, less time and easiness to fabricate a predefined crack are the major advantages of utilizing 3D printing. Applicability of additive manufacturing involves the characterization of the internal defect and to predict the performance of internal defected specimens. The future scope of the research is to inspect different parameters such as PLA pigment, infill rate, printing speed on the mechanical performance of the material. The metallic materials with internal defects need to be investigated.

5. References

[1] Lee, Dongkeon, Takashi Miyoshi, Yasuhiro Takaya, and Taeho Ha. "3D microfabrication of photosensitive resin reinforced with ceramic nanoparticles using LCD microstereolithography." *J. Laser Micro/Nanoeng.* 1 (2006): 142-148.

[2] Parandoush, Pedram, and Dong Lin. "A review on additive manufacturing of polymer-fiber composites." *Composite Structures* 182 (2017): 36-53.

[3] Caffrey, T. "Additive manufacturing and 3D printing state of the industry annual worldwide progress report." *Engineering Management Research* 2, no. 1 (2013): 209-222.

[4] Kruth, J-P., Ming-Chuan Leu, and Terunaga Nakagawa. "Progress in additive manufacturing and rapid prototyping." *Cirp Annals* 47, no. 2 (1998): 525-540.

[5] V. G. Gokhare, D. N. Raut, D. K. Shinde "A Review paper on 3D-Printing Aspects and Various Processes Used in the 3D-Printing" *International Journal of Engineering Research & Technology (IJERT)* 6 (2017): 953-958

[6] N. Turner, Brian, Robert Strong, and Scott A. Gold. "A review of melt extrusion additive manufacturing processes: I. Process design and modeling." *Rapid Prototyping Journal* 20, no. 3 (2014): 192-204.

[7] Kim, Heechang, Eunju Park, Suhyun Kim, Bumsoo Park, Namhun Kim, and Seungchul Lee. "Experimental Study on Mechanical Properties of Single-and Dual-material 3D Printed Products." *Procedia Manufacturing* 10 (2017): 887-897.

[8] Auras, Rafael A., S. Paul Singh, and Jagjit J. Singh. "Evaluation of oriented poly (lactide) polymers vs. existing PET and oriented PS for fresh food service containers." *Packaging technology and science* 18, no. 4 (2005): 207-216.

[9] Letcher, Todd, and Megan Waytashek. "Material property testing of 3d-printed specimen in PLA on an entry-level 3d printer." In *ASME 2014 International Mechanical Engineering Congress and Exposition*, pp. V02AT02A014-V02AT02A014. American Society of Mechanical Engineers, 2014.

[10] Wittbrodt, Ben, and Joshua M. Pearce. "The effects of PLA color on material properties of 3-D printed components." *Additive Manufacturing* 8 (2015): 110-116.
[11] 11.Chacón, J. M., M. A. Caminero, E. García-Plaza, and P. J. Núñez. "Additive manufacturing of PLA structures using fused deposition modelling: effect of process parameters on mechanical properties and their optimal selection." *Materials & Design* 124 (2017): 143-157.
[12] 13. Farbman, Daniel, and Chris McCoy. "Materials Testing of 3D Printed ABS and PLA Samples to Guide Mechanical Design." In *ASME 2016 11th International Manufacturing Science and Engineering Conference*, pp. V002T01A015-V002T01A015. American Society of Mechanical Engineers, 2016.
[13] Kotlinski, Jaroslaw. "Mechanical properties of commercial rapid prototyping materials." *Rapid Prototyping Journal* 20, no. 6 (2014): 499-510.
[14] Lanzotti, Antonio, Marzio Grasso, Gabriele Staiano, and Massimo Martorelli. "The impact of process parameters on mechanical properties of parts fabricated in PLA with an open-source 3-D printer." *Rapid Prototyping Journal* 21, no. 5 (2015): 604-617.
[15] JSM – 6510 SERIES user manuals, 1st ed, JEOL Ltd, 2014.
[16] Torrado, Angel R., and David A. Roberson. "Failure analysis and anisotropy evaluation of 3D-printed tensile test specimens of different geometries and print raster patterns." *Journal of Failure Analysis and Prevention* 16, no. 1 (2016): 154-164.
[17] Maiti, S. K., G. Krishna Kishore, and Mourad, A-H. I.. "Bilinear CTOD/CTOA scheme for characterisation of large range mode I and mixed mode stable crack growth through AISI 4340 steel." *Nuclear Engineering and Design* 238, no. 12 (2008): 3175-3185.
[18] Maiti, S. K., S. Namdeo, and Mourad, A.-H. I. "A scheme for finite element analysis of mode I and mixed mode stable crack growth and a case study with AISI 4340 steel." *Nuclear Engineering and Design* 238, no. 4 (2008): 787-800.
[19] Mourad, A.-H. I., M. J. Alghafri, OA Abu Zeid, and S. K. Maiti. "Experimental investigation on ductile stable crack growth emanating from wire-cut notch in AISI 4340 steel." *Nuclear engineering and design* 235, no. 6 (2005): 637-647.
[20] Ahmed, W. K., and Mourad, A.-H. I. "Fracture assessment of strengthened cracked metallic components using FRP stiffeners." *Mechanics of Composite Materials* 51, no. 3 (2015): 301-312.
[21] Mourad, Abdel-Hamid I. "Pure shear stable crack growth through Compact-Tension-Shear specimen in plane state of stress." *Strength, fracture and complexity* 2, no. 3 (2004): 111-125.

IOP Conf. Series: Materials Science and Engineering **362** (2018) 012027 doi:10.1088/1757-899X/362/1/012027

Experimental Study in Taguchi Method on Surface Quality Predication of HSM

Yan Ji[1,a] and Yueen Li[2,b]

1 PhD., Shandong Jianzhu University, Jinan, P.R,China
2 Asscoiate Professort, Shandong Jianzhu University, Jinan, P.R,China
E-mail: [a] jyan1202@sdjzu.edu.cn, [b]leeyohn@sdjzu.edu.cn

Abstract. Based on the study of ball milling mechanism and machining surface formation mechanism, the formation of high speed ball-end milling surface is a time-varying and cumulative Thermos-mechanical coupling process. The nature of this problem is that the uneven stress field and temperature field affect the machined surface Process, the performance of the processing parameters in the processing interaction in the elastic-plastic materials produced by the elastic recovery and plastic deformation. The surface quality of machining surface is characterized by multivariable nonlinear system. It is still an indispensable and effective method to study the surface quality of high speed ball milling by experiments.

1. Introduction

High-speed milling is a kind of high-efficiency machining method with small cutting depth, high spindle speed and high feeding speed, which can greatly reduce the processing cost and time, and obtain high surface quality. High-speed milling is advanced manufacturing technology, which is one of the hot research fields in the field of science and industry [1-5].

High-speed machining concept from the famous German Dr. Salomon 1924 began a series of experimental research, with its April 1931 published "Death Valley" zone as a symbol of high-speed machining technology was officially born. More than sixty years, people from the cutting mechanism, machine tool system, tool materials, work piece materials, numerical control technology and other aspects of research, to explore effective, practical and reliable high-speed cutting technology.In general, surface roughness or surface accuracy is often used to represent the surface features after metal cutting, so the study of metal surface roughness is the key to study the surface morphology [6,7].

The theoretical research on surface roughness is based on cutting kinematics and tool parameters. However, there are some problems in the accuracy of the model, and there is a big gap with the experimental results, some theoretical results and the measured results difference of 5 times [8]. The main reason is that, in the milling process, in addition to considering the relative movement of the tool-workpiece, the geometric parameters of the tool, the cutting process vibration, tool wear, cutting deformation, cutting heat and other factors on the surface morphology also has a great impact, these The combined effect of factors can not be ignored.

2. HSM Mechanism and Machining Surface Formation Mechanism

According to the Thermos-mechanical coupling theory [9], The plastic deformation, elastic recovery and thermal stress effect of the tool- Making the processing of surface formation process is very complex. In the high-speed ball-end milling process, the work-piece surface is affected by the mechanical load and the thermal load interaction. The effects of surface residual stress, work hardening and surface roughness are as follows:

Because the grain breakage hinder the metal deformation in the cutting process, slip, under the action of oxidation, although the high-speed milling cutting force is not large, but the surface layer has been hardened. At the same time, under the action of thermal load, thermal expansion in the process and the tool to thermal expansion of the role of materials, further exacerbate the degree of surface hardening. The plastic deformation of the material is hindered during the processing, which affects the surface roughness, and the roughness varies with the processing conditions.

2.1 Ball-end milling machining mechanical load

Based on the thermal-mechanical theory , there is a complex process of the work-piece-tool in contact zone, where occurs an overlap between temperature-force coupling field effects. Because of grain squeezing and friction effects, it make the grain distortion, breakage and extension, there is not same deformation among the grains and their boundary, the mechanical load produces the residual stress, the high temperature in machining contact zone make the residual stress complexity. In the ball milling process, the edge line of the tool-tool contact area is affected by the tool position, the tool path and various cutting parameters, so that the length and position of the edge line are different. Which could be explained by the theory of overload elastic recovery, as shown in Fig. 1.

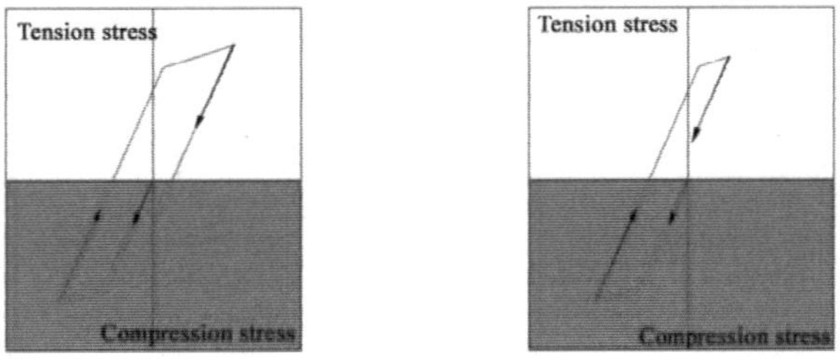

Fig.1 The material residual stress of the surface and the layers

Due to the effect of the periodic external force, when the cutting force is mainly under the pressure, the strain will spring back to produce the residual tensile stress. The load applied by the tool on the surface of the workpiece is reflected by the interaction, so the stress on the surface presents the state of alternating tension and compression.

2.2 Ball-end milling effect of thermal load

Generally speaking, in the high-speed cutting, the metal of the cutting layer is subjected to high-speed friction, slip shearing and squeezing while the local temperature of the knife-work contact area rises gradually, because the surface temperature of the workpiece is much higher than the inner layer The surface layer of local thermal expansion, but by the inner layer of hot metal is not bound, then the surface layer of metal thermal stress, the surface layer to form a temperature gradient, resulting in compression plastic deformation.

3. The Effect of High Speed Milling Parameters on Surface Quality

Table 1. $L_{25}(5^6)$ ball-end milling cutter high-speed milling orthogonal experiment cutting parameters

No.	n(r/min)	f_z(mm/z)	a_e(mm)	a_p(mm)	β_n (°)	β_f (°)	No.	n(r/min)	f_z(mm/z)	a_e(mm)	a_p(mm)	β_n (°)	β_f (°)
1	7000	0.05	0.20	0.20	0	0	13	9000	0.10	0.05	0.15	30	0
2	7000	0.08	0.15	0.15	10	10	14	9000	0.15	0.20	0.10	40	10
3	7000	0.10	0.10	0.10	20	20	15	9000	0.20	0.15	0.08	0	20
4	7000	0.15	0.08	0.08	30	30	16	10000	0.05	0.08	0.15	40	20
5	7000	0.20	0.05	0.05	40	40	17	10000	0.08	0.05	0.10	0	30
6	8000	0.05	0.15	0.10	30	40	18	10000	0.10	0.20	0.08	10	40
7	8000	0.08	0.10	0.08	40	0	19	10000	0.15	0.15	0.05	20	0
8	8000	0.10	0.08	0.05	0	10	20	10000	0.20	0.10	0.20	30	10

9	8000	0.15	0.05	0.20	10	20	21	11000	0.05	0.08	0.08	20	10
10	8000	0.20	0.20	0.15	20	30	22	11000	0.08	0.05	0.05	30	20
11	9000	0.05	0.10	0.05	10	30	23	11000	0.10	0.15	0.20	40	30
12	9000	0.08	0.08	0.20	20	40	24	11000	0.15	0.10	0.15	0	40
							25	11000	0.20	0.08	0.10	10	0

The influence factors of high speed milling include: cutting parameters, tool parameters, control system parameters, machine performance parameters, and work-piece material parameters. Orthogonal experiment is a mature multi-parameter analysis method. In the experiment, the feed rate per tooth, axial depth of cut, radial cutting width, inclination angle, side angle and rotation speed are taken as machining parameters. Surface roughness and work hardening as the criterion of surface quality.

The influence degree of various processing parameters on the surface quality of high speed ball milling was analyzed, and the influence of processing parameters on the surface quality was analyzed.The high-speed machining experiment is carried out in DMU-70V high-speed 5-axis CNC . The work-piece material is hot work die steel H13 (54-56HRC). The tool is solid carbide tool with diameter of 10mm and 2 teeth.The experimental parameters were processed using a standard 6-factor, 5-level, 25-group orthogonal experiment table. The parameters were set as shown in Table 1.

After the end of the high-speed milling experiment, using white light interferometer, hardness tester and other equipment on the test surface, surface roughness and hardness measurements shown in Table 2:

Table 2. The roughness and hardness test results

No.	Ra(μm)	HL	No.	Ra(μm)	HL
1	1.739	215	13	0.357	229
2	0.279	235	14	0.841	232
3	0.317	230	15	0.495	212
4	0.307	226	16	0.183	235
5	0.595	203	17	0.361	250
6	0.333	235	18	0.447	226
7	0.395	239	19	0.412	245
8	0.191	222	20	0.784	244
9	0.343	238	21	0.22	225
10	0.736	233	22	0.2	243
11	0.22	238	23	0.571	245
12	0.509	240	24	0.243	243
			25	0.296	231

4. High-speed Milling Tool Position on the Impact of Surface Quality

The influence of the tool posture on the surface quality in high speed milling is various, and the effect on the machined surface topography and the work hardening is also more direct. In this section, the workpiece is machined with high speed machining.

Table 3. surface roughness, surface hardness test results

β_f (°)	Ra(μm)	HL	Ra(μm)	HL	Ra(μm)	HL
	f_z=0.1(mm/z)		f_z=0.2(mm/z)		f_z=0.3(mm/z)	
0	0.756	103	0.669	105	2.823	95
5	1.132	100	0.849	108	0.468	99
10	1.142	119	0.953	116	0.617	101
15	1.065	110	1.108	105	0.546	113
20	1.521	100	1.44	103	0.521	107
25	1.583	118	1.307	109	0.618	108
30	1.836	115	1.313	102	0.671	112
35	1.471	107	1.195	103	0.684	103
40	1.718	92	1.309	92	0.619	94

45	1.16	85	0.968	109	1.03	85

4.1 High Speed Ball Milling Experiment

The experiment is the same as above, and the processing is the same. The purpose of the experiment is to measure the roughness and surface hardness of different parameters by changing the position of the tool under steady-state cutting condition, so as to reflect the influence of cutting parameters on roughness.

4.2Effect of tool posture on roughness

Fig.2 shows the relationship between the tool posture and the surface roughness. The influence of the inclination angle on the machined surface is not particularly obvious under the condition that the other machining parameters are constant.The surface roughness value increases with the increase of feed per tooth. With the increase of inclination angle, the surface roughness value increases slowly. When fz = 0.1mm / z, When the angle of inclination is 0 °, the surface roughness value is 2.823um, which indicates that the surface quality is poor at this time, and the surface roughness value shows an increasing trend when the inclination angle changes from 5 ° to 45 °.

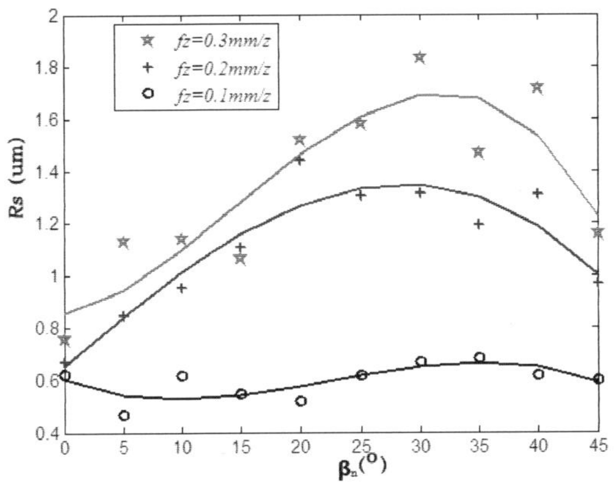

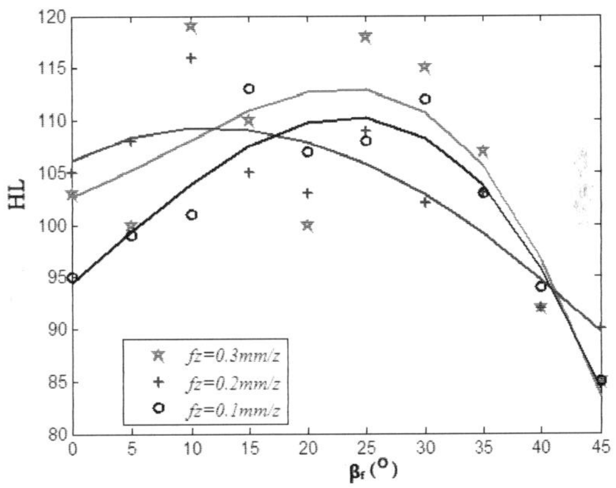

Fig.2 Effect of tool pose on Surface Roughness Fig.3 Effect of surface hardness of the tool posture

Further analysis of the experimental results, the establishment of surface roughness and rake angle polynomial fitting formula:

$$Ra = a \cdot \beta_f^{\,3} + b \cdot \beta_f^{\,2} + c \cdot \beta_f + d \quad (1)$$

Where a, b, c and d are polynomial fitting coefficients, and Table 4-5 shows polynomial fitting coefficients for different tooth feed fz (mm / z)

Table 4. Roughness Ra (μm) Polynomial Fit Factor

	f_z (mm/z)	*a*	*b*	*c*	*d*
Ra	0.3	0	0.0019	0.0097	0.8547
	0.2	0	-0.0002	0.0392	0.654
	0.1	-0.0001	0.0113	-0.2768	2.3815

From the analysis of the characteristics of milling, different inclination angle corresponding to different cutting edge, involved in the cutting edge of the material caused by plastic deformation of the surface morphology of the difference.

4.3 Impact of tool posture on surface hardness

Different inclination angle produces different cutting force and cutting heat, especially the change of the arbor, the depth of the hardened layer and the hardness have great changes. The surface hardness is affected by the elastic recovery of the material, and the different tool posture hinders the ability of the surface elastic recovery. Fig.3 shows the relationship between tool posture and surface hardness.

According to the experimental data, the polynomial fitting coefficient between the surface hardness as shown in Table 5:

Table 5 surface hardness HL polynomial fitting coefficient

	f_z (mm/z)	*a*	*b*	*c*	*d*
HL	0.3	-0.0009	0.0229	0.4169	102.6308
	0.2	0.0013	-0.0917	1.4073	104.6965
	0.1	-0.0006	0.0007	1.0014	94.4448

From the analysis of Fig.3: fz = 0.3mm / z, 0.2mm / z, 0.1mm / z, with the inclination from 0 ° ~ 45 ° increase, the highest surface hardness difference, but the numerical range , And the hardness varied from 85 to 120.When fz = 0.3mm / z and 0.2mm / z, the peak appeared at the inclination angle of 10 ° and 25 °. And an inclination angle of 15 ° and 30 °.By analyzing the work hardening trend, it is found that the variation of the hardness of the tool is obvious, and its change rule is basically the same as that of the tool posture. The reason is mainly due to the accumulation effect of the plastic deformation in the feed direction. Residual stress has the same cumulative effect.

5. Conclusions

The effect of processing parameters on the surface quality was studied in this paper from the experimental point of high speed ball milling. The effect of processing parameters on the surface quality was analyzed according to the orthogonal experiment. The influence of processing parameters on surface quality was analyzed. According to the influence characteristics of the tool posture, the rule of the surface roughness and the surface hardness of the tool posture are studied. The polynomial fitting formula is given, which provides the theoretical basis for the surface quality prediction.

6. References

[1] Xiao, J. M.. Modeling of surface roughness with ball-end cutters milling based on scallop height and optimum of cutting parameters. Modular Machine Tool & Automatic Manufacturing Technique,2015.

[2] Daymi, A., Boujelbene, M., Linares, J. M., & Bayraktar, E.Influence of workpiece inclination angle on the surface roughness in ball end milling of the titanium alloy ti-6al-4v. Publishers Weekly, 86(4), (2009) 1028-1029.

[3] Chen, J. S., Huang, Y. K., & Chen, M. S. A study of the surface scallop generating mechanism in the ball-end milling process. International Journal of Machine Tools & Manufacture, 45(9), (2005)1077-1084.

[4] Fard, M. J. B., & Bordatchev, E. V.Experimental study of the effect of tool orientation in five-axis micro-milling of brass using ball-end mills. The International Journal of Advanced Manufacturing Technology, 67(5), (2013)1079-1089.

[5] Han, S. G., & Zhao, J. Effect of tool inclination angle on surface quality in 5-axis ball-end milling. Advanced Materials Research, 97-101, (2010)2080-2084.

[6] Sonawane, H. A., & Joshi, S. S. (2015). Modeling of machined surface quality in high-speed ball-end milling of inconel-718 thin cantilevers. The International Journal of Advanced Manufacturing Technology,78(9), (2010)1751-1768.

[7] Wong, D. (2007). Particleboard simulation model to improve machined surface quality. (Doctoral dissertation).

[8] Wang, S. J., To, S., Chen, X., Chen, X. D., & Ouyang, X. B. An integrated optimization of cutting parameters and tool path generation in ultraprecision raster milling. The International Journal of Advanced Manufacturing Technology, 75(9), (2014)1711-1721.

[9] Rao, R. V. Modeling and optimization of machining processes. (2011) 55-175.

[10] Chen, X. X., Zhao, J., Li, Y. E., Han, S. G., Cao, Q. Y., & Li, A. H. Numerical modeling of high-speed ball end milling with cutter inclination angle. part b j eng manuf. Proceedings of the Institution of Mechanical Engineers Part B Journal of Engineering Manufacture, 226(4), (2011)606-616.

IOP Conf. Series: Materials Science and Engineering **362** (2018) 012028 doi:10.1088/1757-899X/362/1/012028

Reliability Modeling of Double Beam Bridge Crane

HAN ZHU[1, a], YIFEI TONG[1, b], JIAHUI luan[1, c], XIANGDONG LI[2, d*]

1 Nanjing University of Science and Technology, Nanjing, China
2 Jiangsu Province Special Equipment Safety Supervision Inspection Institute, Nanjing, China
E-mail: [a]1019632618@qq.com, [b]tyf51129@aliyun.com,
[c]484503036@qq.com, [d*]765801038@qq.com.

Abstract. This paper briefly described the structure of double beam bridge crane and the basic parameters of double beam bridge crane are defined. According to the structure and system division of double beam bridge crane, the reliability architecture of double beam bridge crane system is proposed, and the reliability mathematical model is constructed.

1. Introduction

Bridge crane is a large lifting equipment and widely used in business workshop, which can effectively reduce the labor intensity and improve work efficiency [1]. However, most of the bridge cranes currently used in our country are produced before the 1980s. Considering the cost of production, only a small amount of bridge crane was scrapped. Most of the bridge cranes entering the aging state are still in service. The aging problem is increasingly prominent and bringing greater risks. The key to solving the above problems lies in establishment of a set of effective fault detection and diagnosis system and reasonable risks assessment. The reliability analysis of bridge cranes is an important means to ensure safe and reliable operation of cranes and is also an important step to establish the fault detection and diagnosis system. In this paper, a double beam bridge crane is taken as the research object, whose reliability model is established and reliability parameters are determined.

2. Double beam bridge crane system structure

The double beam bridge crane (shown in Figure 1.) is composed of three parts: metal structure, mechanical transmission and electrical equipment.

Figure 1. Double beam bridge crane.

(1) Metal structure includes bridge, trolley frame and control room. The bridge is a removable metal structure consisting of main beam, end beam, trolley rails and guardrail. The trolley frame is welded by a steel plate with lifting mechanism and car movement organization. Control room is suspended under the bridge and equipped with electrical control equipment for the operator to use.

(2) In order to realize the different requirements of crane, the mechanical transmission is divided into lifting mechanism, trolley traveling mechanism and cart traveling mechanism. The lifting mechanism used to realize the lifting of the goods and installed on the upper of the trolley frame, is composed of motor, brake, reducer, coupling, reel and pulley. The trolley traveling mechanism drives the load for lateral movement, which is composed of motor, brake, coupling, reducer, wheel and drive shaft and other components. The cart traveling mechanism drives cart wheels running along the track and the trolley and goods for longitudinal movement, composed of wheel, angle shaft, reducer, coupling, drive shaft, brake and motor and other components.

(3) The electrical part consists of electrical circuit and electrical equipment.

3. The main parameters of double beam bridge crane

The following are the main parameters of the double beam bridge crane:

(1) Lifting capacity: 5T 10T 16/3.2T
20/5T 32/5T 50/10T

(2) Span: 10.5M 13.5M 16.5M 19.5M
22.5M 25.5M 28.5M 31.5M

(3) Working conditions:

A5 (Used for less frequent jobs such as general machining and assembly workshops).

A6 (Used for more frequent jobs, such as auxiliary lifting in metallurgy and foundry shops).

A7 (For busy use and hot metal lifting).

When the weight is expressed as a fraction, the numerator indicates the main lifting weight, and the denominator indicates the secondary lifting weight.

A crane with a weight of 5 or 10 tons is a single hook crane with only one set of lifting mechanism. 16/3.2-50/10 tons of cranes have two hooks, so there are two independent lifting mechanisms. The main hook is used to lift heavy objects. In addition to lifting lighter objects, the auxiliary hook can also be used in conjunction with the main hook to turn or reverse the object in its rated load range. But it must be noted that two hooks are not allowed to simultaneously ascend two objects. When two hooks work together, the weight of the object must not exceed the weight of the main hook.

4. Selection of reliability parameters of double beam bridge crane

In engineering, the following three reliability parameters are commonly used.

1) Reliability

Reliability refers to the probability of the product completing the set function under the specified conditions and within the specified time.

2) Failure rate [2]

Failure rate refers to the probability that a product has not been malfunction at a certain time and the failure occurs within the unit time after this time.

3) Mean time between failures (MTBF) [3]

MTBF is an important reliability parameter for repairable products. The repairable product occurs N_0 failures in use, after each repair to continue to use, the continuous working time (i.e. the time between failures) is $t_1, t_2, \cdots, t_i, \cdots, t_{N_0}$, and the observed value of MTBF is

$$MTBF = \frac{1}{N_0}\sum_{i=1}^{N_0} t_i = \frac{T}{N_0} \tag{1}$$

where, T is the total working time of the product.

Among them, the application scope of MTBF is extensive, and it is also a measure of the basic reliability of the repairable product, which reflects the related requirements of the product on maintenance resources [4]. For a repairable large mechanical equipment of double beam bridge crane, because the reliability of its parts and components have big difference, MTBF is selected as its reliability design and evaluation of reliability parameters.

5. Establishment of reliability model for double beam bridge crane system

When establishing the reliability model, we need to collate and analyze the failure information of the crane, and judge the source of the fault, so as to find the mechanism of the failure and deal with the fault in time. Combined with the structure of double beam bridge crane, mechanical transmission is the most important part of crane and the most prone to failure [5]. Therefore, the double beam bridge crane are divided into three important subsystems.

1) Lifting mechanism system

The lifting mechanism includes motor, brake, reducer, coupling, reel and pulley group etc.

2) Trolley traveling mechanism system

The trolley traveling includes motor, coupling, reducer, brake, drive shaft and wheel etc.

3) Cart traveling mechanism system

The cart traveling mechanism includes motor, angle shaft, reducer, coupling, drive shaft, brake and wheel etc.

It can be seen that the double beam bridge crane consists of three subsystems, and its reliability system can be made up of a simplified series model. The reliability diagram of the double girder bridge crane subsystem is shown in Figure 2.

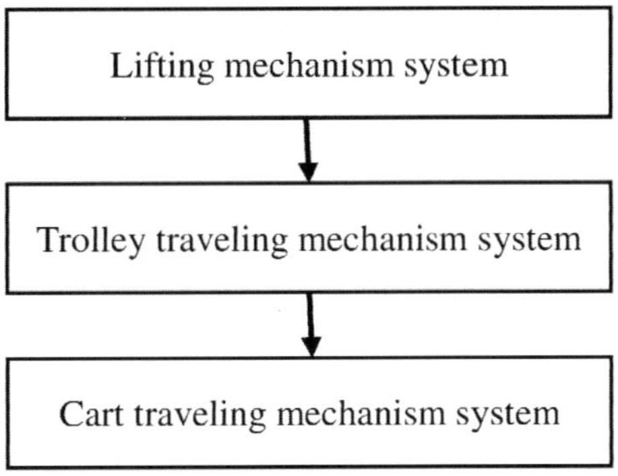

Figure 2. Reliability architecture of the subsystem of double beam bridge crane.

According to the above analysis, the reliability model of double beam bridge crane system is as follows:

$$R(t) = R_L(t) \times R_T(t) \times R_C(t) \tag{2}$$

where, $R(t)$ is reliability of double beam bridge crane system, $R_L(t)$ is reliability of lifting mechanism system, $R_T(t)$ is reliability of trolley traveling mechanism system, $R_C(t)$ is reliability of cart traveling mechanism system.

The reliability block diagram of the double beam bridge crane obtained by connecting the parts of each subsystem is shown in figure 3.

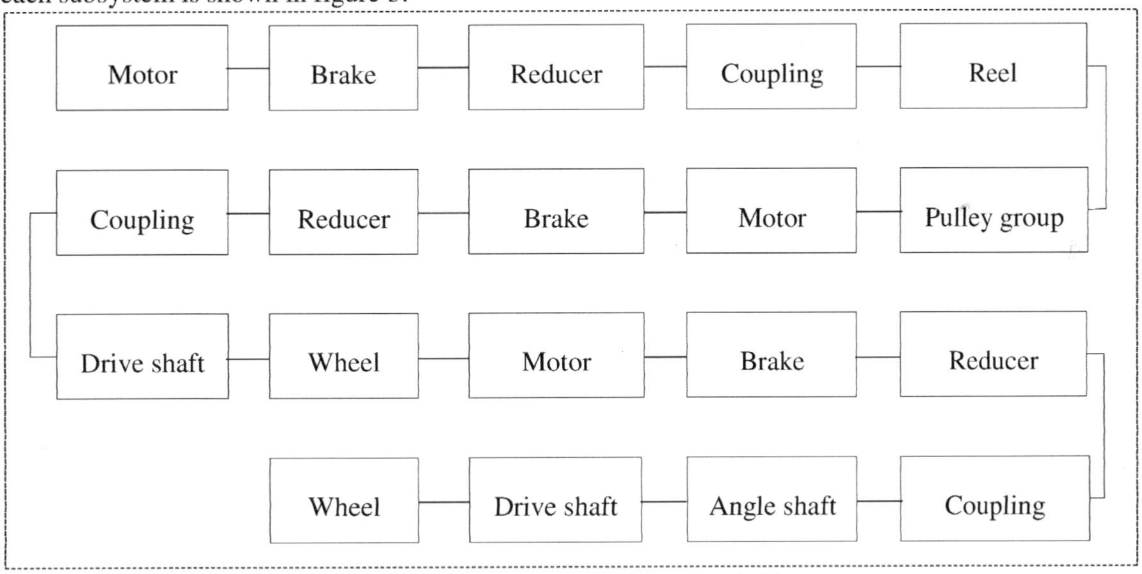

Figure 3. Reliability architecture of double beam bridge crane system.

For double beam bridge crane, the subsystem is also composed of various component. The reliability model of the subsystem is as follows:

$$R_S(t) = R_1(t) \times R_2(t) \times \cdots \times R_n(t) \tag{3}$$

where, $R_S(t)$ is reliability of the crane subsystem, $R_i(t)$ ($i = 1,2,\cdots,n$) is reliability of part i.

6. Conclusion

According to the characteristics of double beam bridge crane, the system is divided into three subsystems. The reliability architecture is proposed as well as its reliability model.

7. References

[1] Wei Ye,Yifei Tong, Dongbo Li, Xiangdong Li, Xucheng Yuan, Research on bridge cranes energy-saving index reduction based on rough set theory, Telkomnika Indonesian Journal of Electrical Engineering. 11(2013) 7269-7276.

[2] Barlow R E, Proschan F, Statistical theory of reliability and life testing: probability models, Florida State Univ Tallahassee, 1975.

[3] Yamamoto H, Kitamura K, Yokota M, Hayashi R, Uematsu Y, Availability improvement schemes for multi-carrier optical transmission systems, IEICE Communications Express. 6(2017) 633-638.

[4] Wang Z, Reliability maintainability indemnificatory requirement demonstration, National Defense Industry Press, Beijing, 2011, pp. 49-56.

[5] Wu F Q, The Dynamic Strength Calculation on the Reel Fulcrum of Crane Based on ADAMS and FEA, Applied Mechanics & Materials. 599-601(2014) 551-554.

Acknowledgments

This work was financially supported by the Program of Science Foundation of General Administration of Quality Supervision and Inspection of Jiangsu Province (KJ175940). The supports are gratefully acknowledged.

Institute of Physics
Dirac House, Temple Back
Bristol BS1 6BE UK

ISSN: 1757-8981
ISBN 978-1-5108-6341-5
9 781510 863415

CURRAN ASSOCIATES INC.
proceedings.com

3rd International Conference on Manufacturing, Material and Metallurgical Engineering (ICMMME 2018)

IOP Conference Series: Materials Science and Engineering Volume 409

Kuala Lumpur, Malaysia
17 – 19 March 2018

IOP
Institute of Physics